The CIVIL WAR HISTORY Series

USS
CONSTELLATION

AN ILLUSTRATED HISTORY

On July 2, 1999, the 34-star flag returned to her rightful place on board the sloop of war USS *Constellation*. As she "showed the flag" while patrolling the Mediterranean to assure the Old World of the steadfastness of the United States, the *Constellation* again displays the stars and stripes to give inspiration for the future through the history of the past.

THE CIVIL WAR HISTORY SERIES

USS CONSTELLATION

AN ILLUSTRATED HISTORY

STEPHEN R. BOCKMILLER
AND LAWRENCE J. BOPP

ARCADIA
PUBLISHING

Published by Arcadia Publishing
Charleston, South Carolina

Printed in the United States of America

Library of Congress Catalog Card Number: 00-103694

For all general information contact Arcadia Publishing at:
Telephone 843-853-2070
Fax 843-853-0044
E-mail sales@arcadiapublishing.com
For customer service and orders:
Toll-Free 1-888-313-2665

Visit us on the Internet at www.arcadiapublishing.com

In times of war and peace, prosperity and despair,
our course was always steadied by the firm hand of men who cared.

We lovingly dedicate this book to our fathers,
Robert R. Bockmiller and Ambrose L. Bopp Jr.,
the helmsmen of the ships that are our lives.

On the cover: Trainees at the U.S. Naval Training Station at Newport, RI, man the yards of the
USS *Constellation* in 1912.

CONTENTS

Acknowledgments

In the process of writing this book and locating the necessary illustrations, we became painfully aware that, while one or two names are listed on the byline of a book, rarely would such a project be completed without the assistance, guidance, and support of a legion of friends behind the scenes.

We thank the following individuals for their assistance in this project: Larry Clemens and Mary Catalfano, The Adm. Chester W. Nimitz Library, U.S. Naval Academy, Annapolis, MD; Jennifer Hafner, The Maryland State Archives, Annapolis, MD; M.E. Warren, Annapolis, MD; Charles R. Hazard, Owings Mills, MD; John V. Quarstein, The Virginia War Museum, Newport News, VA; J. Michael Moore, The Lee Hall Mansion, Newport News, VA; David L. Sullivan, Rutland, MA; Werner Hirsch, New Haven, CT; Richard Berglund, Silver Spring, MD; Ross Kelbaugh, Michael White, Christopher Hentz and Charles Klein, all of Baltimore, MD; Jack Green, The U.S.N. Historical Center, Washington, D.C.; Anthony Nicolosi and Evelyn M. Cherpak, The U.S. Naval War College Museum, Newport, RI; Katie A. Bopp of Baltimore, MD; and Stefania G. Good of Elizabethtown, PA.

Last, but by no means least, we thank the staff of the USS *Constellation*, Baltimore, MD, for their ongoing support of this project and their stewardship of the *Constellation*, including Christopher Rowsom, Executive Director; Glenn Williams, Exhibit Curator and Historian; Paul Powichroski, Ship's Manager; Christy Schmitt, former Administrative Assistant; and Louis F. Linden, former Executive Director. Almost half of the images in this book are from the ship's archives, and this project would not have been possible without their cooperation and support.

Photo Credits

Unless credited below, all images used in this book are from the archives of the USS *Constellation*. Images are noted by page number with "a" for the first image and "b" for the second image.

Richard Berglund: 83a; Stephen R. Bockmiller: 17b, 47a, 70a, 70b, 71b, 72a, 73a, 73b, 75a, 75b, 76a, 76b, 78a, 78b, 81a, 81b, 85a, 95b, 97b, 122a; Lawrence J. Bopp: 2, 15b, 31b, 117, 118a, 118b, 119a, 119b, 120a, 120b, 121a, 121b, 122b, 123a, 123b, 124a, 124b, 125, 128b; The Boston National Historical Park: 17a; The Civil War Library of Philadelphia: 37; The Connecticut State Library: 25b; The Connecticut Valley Historical Museum: 19; Arthur N. Disney Sr. (courtesy of the USN Historical Center): 21, 44; W.A. Green, *The Providence Plantations for 250 Years* (1886): 65; Charles R. Hazard (with minor alterations by Charles Klein): 10, 11a; Christopher Hentz: 15a, 35, 126a, 126b, 127a, 128a; Werner Hirsch: 74a, 74b, 79a, 79b; Ross Kelbaugh: 86a; The Library of Congress: 24a, 29, 30a; Massachusetts Commandery, Maine Society, Order of Founders and Patriots, and the U.S. Army Military History Institute: 34; Maine Commandery, MOLLUS: 32; The Maryland State Archives-Charlie Dell Photographic Collection: 64; The Adm. Chester W. Nimitz Library, U.S. Naval Academy: 36b, 43b, 48a, 48b, 49a, 49b, 50a, 52, 53, 55a, 56a, 83b, 93a; Russel Parks: 18b; David Dixon Porter, *The Naval History of the Civil War*: 16, 22, 25a, 47b, 56b, 58a; John Moran Quarstein: 12a; David L. Sullivan: 13b, 45; The U.S. Naval Historical Center: 11b, 12b, 18a, 23b, 24b, 26, 27, 28, 31a, 33, 41, 42, 43a, 46, 51, 54a, 54b, 57, 60a, 61a, 61b, 69, 77, 80b, 90a, 90b, 94b, 95a, 96b; The U.S. Naval War College Museum: 66a, 66b, 67a, 71a, 72b, 84a, 84b, 98a; Official U.S. Navy Photograph: 93a, 99, 127b; The U.S. National Archives: 68a, 68b; The U.S. Naval Institute: 9; Lee A. Wallace Jr.: 23a; M.E. Warren: 50b, 55b; and Michael D. White: 30b, 122a inset.

INTRODUCTION

To help give order to their world, the ancient Greeks devised images called constellations by connecting stars in the heavens to form familiar images. Likewise, the Congress of the young republic known as the United States of America ordered that its flag contain a constellation of 13 stars. Unfortunately, the particular constellation to be used was not defined and numerous variations of stars appeared on early American flags. The true design was hidden in vague definition.

Such is the history of the ship moored today in Baltimore Harbor known as the USS *Constellation*. Despite being shrouded in controversy since her construction in 1854, this last surviving ship afloat from the Civil War era has managed to serve and educate her citizenry for a time spanning nearly 150 years. Although she is not the original frigate constructed in 1797, she has managed to carry on the proud title of her namesake into the modern age and a new millennium.

The remarkable history of this ship is revealed within these pages through the collection of numerous images documenting not only the ship, but also the people, places, and events that were effected by her presence. From her disguised creation near Norfolk, VA, the story is told of her noble involvement in suppressing the African slave trade. Her service to her glorious Union is borne out in relating her efforts during the Civil War along with the story of the men comprising her crew. Her ties with education are highlighted in relating her use as a training ship for both the United States Naval Academy and then for the Naval Training Station at Newport, RI. Finally, the story of her restorations, both false and true, brings her full circle back to her origin.

By focusing on the true story of her past, the controversy surrounding her origin is resolved. Fact rather than opinion is valued in recounting her history within these pages with hopes that false allegations about the ship will finally be put to rest. The story will tell how she is <u>not</u> the original frigate; she was not a slave ship but rather served to stop this cruel trade; and that she was not destroyed but remains intact today within a protective shell.

Ever since the centennial of the Civil War, the popularity of that era in our nation's history has enabled its story to be told and retold through the media of print, radio, film, and television. Thousands of Americans have become reenactors performing battle scenarios for the public, dramatizing the story of this war. Along the way, their efforts have renewed interest in America's past and preserved innumerable battlefields, buildings, and artifacts. Despite the fact that none of the land action would have been successful without the effort and support of the United States Navy, this branch of the service has received little attention in relating the story of the conflict. Thus, with the restoration of the *Constellation* as the sole surviving Civil War vessel afloat, this one ship becomes more important now than ever. Along with recounting her own story of adventures in the 19th and 20th centuries, she must now take on her most demanding new mission in the 21st century, to represent and to relate the entire naval history of the Civil War. The purpose of this book is to not only share her story with the reader, but to assist her in fulfilling her mission.

Using over two hundred images chosen from the vast archives of her past, her story is brought forth in vivid detail. As each line is drawn to explain the connection of the stars that make up a constellation from the dark of a night sky, each image helps the reader focus upon each individual piece of history that is being brought forth from the murky darkness of the past. Her image becomes more and more defined as her story unfolds until the full brilliance of her creation and restoration today culminates with her present educational mission. Her stars will burn ever brighter as she brings light from the nation's past to show the way to future glory in stories yet untold.

The stories of tens of thousands, some still living today, remain to be told. Whether they be any one of the hundreds of slaves liberated by the *Constellation*, to a novice "naval cadet" from the U.S. Naval Academy, to a member of her Marine Guard in the 1880s, to the sailor operating the ship's radio when the message was received from Hawaii of the Japanese attack on December 7, 1941, the *Constellation* has left her indelible mark on the memories of far too numerous and diverse a population to estimate. We hope her story as told here serves to widen the population affected by her memory and inspires you to protect and ensure her future.

One

CONNECTING THE STARS

In 1844, just three years shy of her 50th birthday, the old war-weary *Constellation* slowly made her way into Gosport Navy Yard in Portsmouth, VA. She was on her way under orders to be put once again "in ordinary," not knowing that this would be for the last time. Her weathered timbers mirrored the tales of her past glory in serving her country. She was one of the first six ships authorized by Congress in 1794 to comprise the new United States Navy. Her name was derived from the "constellation of stars" authorized by Congress in their description of the official flag of the United States. She won her first victory at sea by defeating the French ship *L'Insurgent* during the Quasi-War on February 9, 1799. Just a year later, she added to Navy laurels by defeating another French ship, *La Vengeance*. Scars from cannonballs and small arms fire from her service during the Barbary Wars were also visible reminders of conflicts in which she had taken part.

While being repaired at Norfolk during the War of 1812, she was bottled up in the harbor by a British blockade. Thus, the Navy was deprived of her service during this time of conflict. The next 30 years would see her travel the globe, showing the flag in defense of free seas for American shipping. She was dispatched to South America to protect American interests, fought the last of the Caribbean pirates, patrolled the Mediterranean, and participated in the efforts of the American and British Navies to stop the illegal shipment of slaves from Africa. Her last mission of 16 months' duration had taken a hard toll as she had sailed over 48,000 miles. The gallant lady was now returning home for a much needed rest.

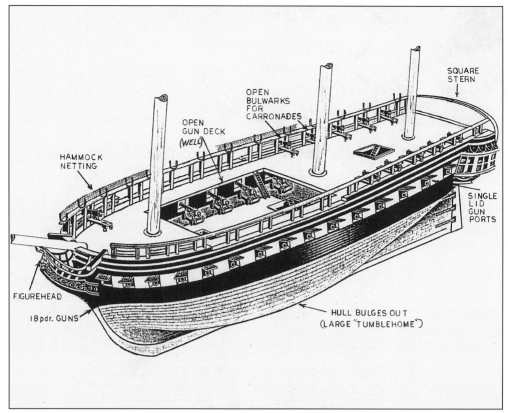

Her rest, however, turned into a death of sorts. After languishing at Gosport for seven years, her fate was debated in Congress. Numerous arguments to repair her, reconfigure her, or even convert her to steam power all came to nought. By 1853, the Navy officially condemned her and ordered her to be dismantled. She was to have a "resurrection," however, as the Navy lobbied to have a new ship constructed in her place.

Working under the subterfuge of "repairs," the Navy actually began building a new ship about 900 yards from where the original *Constellation* was being dismantled. Thus, unwittingly, the Navy itself would originate the arguments about the authenticity of the *Constellation*. To further compound the argument, some salvageable timbers from the original ship, particularly the ship's knees, were used in constructing the new vessel.

The former 38-gun frigate was 164 feet in length and 41 feet wide. Her spar deck opened to a large "well" to allow for ventilation of the gun deck below. Her spar deck carried short range cannons, mostly used as anti-personnel weapons, called carronades, which required that openings be cut into her bulwarks to allow them to be fired. Her gun deck used single piece covers for the gun ports and no galley cover existed on the frigate. Her stern was squared and her sides had more of a pronounced curve commonly called a "tumble home." All of these factors, and more, combined to give her the distinct configuration of an 18th-century frigate, very much unlike the new ship that was being built using the same name. The typical crew of the early frigates of the United States Navy totaled approximately 340 men.

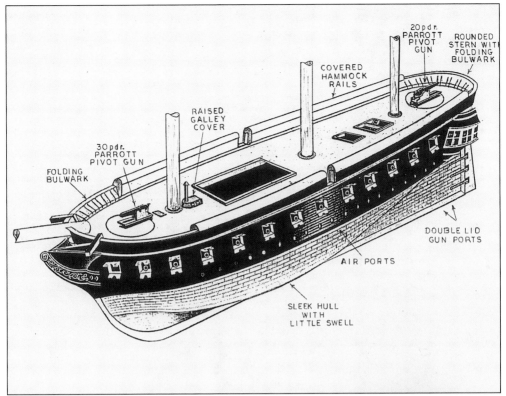

The second *Constellation*, the ship that still exists today, was a far sleeker ship designed by naval architect John Lenthall. She was made 12 feet longer than the frigate, making her 176 feet in length. Her beam (width) was 42 feet and she had a rounded stern that was much stronger than the former square transom. Her spar deck was almost completely closed with a small hatch opening, rather than the large well of the frigate. Her sides had less of a tumble home than her predecessor, which allowed her greater speed.

The greatest difference was in her armament. She was to carry sixteen 8-inch shell guns and four 32-pounders as broadside guns. Later, for her Civil War service, her spar deck was modified to carry two deck pivot guns: a 30-pounder Parrott Rifle forward and a 20-pounder Parrott Rifle aft. Iron rails on deck coupled with folding bulwarks made of wood with iron hinges allowed these guns to be turned and fired. She also had 36 portholes added for ventilation of the berth deck. Her crew was reduced to require approximately 280 men.

John Lenthall (pictured here) was the chief of the U.S. Navy Bureau of Construction, Equipment and Repair. His plans for the *Constellation* were a precursor to the ideas he would later formulate to incorporate the use of steam in future naval vessels. The *Constellation* was the very last all-sail ship designed and constructed by the Navy. Fortunately, a copy of Lenthall's plans for this ship were preserved in Lenthall's papers, as the originals were destroyed when the Gosport Navy Yard was burned by the Navy upon its evacuation at the beginning of the Civil War. The yard was put to the torch to thwart Confederate plans to make use of the yard, its ships, and its machinery for the creation of their own navy.

The Gosport Navy Yard at Portsmouth was one of the largest in the United States and was looked upon with envy by foreign nations. It contained a huge granite dry dock capable of repairing even the huge "ships-of-the-line." Experimental projects were developed here, particularly in the testing of iron plate against cannon fire. It was in this "state-of-the-art" yard that the USS *Constellation* would see her rebirth as a sloop of war. After two years of construction, the *Constellation* was put in service with Capt. Charles H. Bell commanding. On July 28, 1855, the ship was officially back in commission. After several sea trials, Captain Bell took her on her initial journey to patrol the Mediterranean Sea. Her orders were to protect American merchant vessels and to "show the flag" to discourage any European or North African country from interfering with the United States. Her patrol ended in 1858 when she was ordered to Boston and decommissioned.

Captain Bell would be the first of many officers who served on the *Constellation* who would eventually become an admiral. A midshipman in the War of 1812, he served aboard the USS *Macedonian*. In 1824 and 1825, Lieutenant Bell was given his first command: the USS *Ferret*. Ten years later, he was the executive officer of the USS *Vincennes*, flagship of the Pacific Squadron. Bell served several years in the African Squadron in the 1840s before promotion to the rank of captain in 1854 and transfer to the *Constellation* in 1855. He next served as the commandant of the navy yard and naval station at Norfolk, VA, until August 1860. When the war erupted, he served on the Board to Better Organize the Navy in late 1861. He is shown here as a rear admiral, to which he was elevated in 1862. He led the Pacific Squadron from the USS *Lancaster* until 1864. Adm. Charles Bell retired as commandant of the New York Navy Yard in 1868 and died in 1874 at age 77.

Italian artist Tomasso di Simone was commissioned by Captain Bell to make three paintings of the ship. This painting was completed in 1856 while the ship sat in the port of Naples, Italy. Her sleek lines are well displayed here but, for reasons unknown, the artist failed to show the portholes that ventilated the berth deck.

In charge of the Marines on board during her initial cruise was Lt. Thomas Field, pictured here in his Civil War–era frock. Field was brevetted first lieutenant on September 13, 1847, for meritorious conduct in the Battle of Mexico City. He received his full rank on October 15, 1854, prior to sailing on the *Constellation*. During the Civil War, he moved from captain to major as he distinguished himself on board the USS *Saranac*. He went on to command the barracks at Philadelphia, Norfolk, and Washington Navy Yards. He was honored to be chosen by Commandant Jacob Zeilen as the Marine Corps representative to accompany the body of Abraham Lincoln to Springfield. In later years, his reputation suffered when he was found guilty of scandalous conduct and disobedience of a direct order of the secretary of the navy. In spite of his troubles, including a two-year suspension from rank and command, he retired a full colonel in 1889. Ironically, he died at Wayne, PA, in 1905 on February 12—Lincoln's birthday.

The four deck plans of the *Constellation* found on the following page reveal the basic layout of the ship. As one descends to each successive lower deck, the overhead (or ceiling) gets lower and lower. To the novice, it appears that only short men could be hired to work aboard such a ship. However, in reality, the average height of the crew was about 5'10" tall. The closer overhead on each lower deck provided a low center of gravity to stabilize the ship as a floating gun platform. People could always adapt to their environment but the guns needed calculated stability.

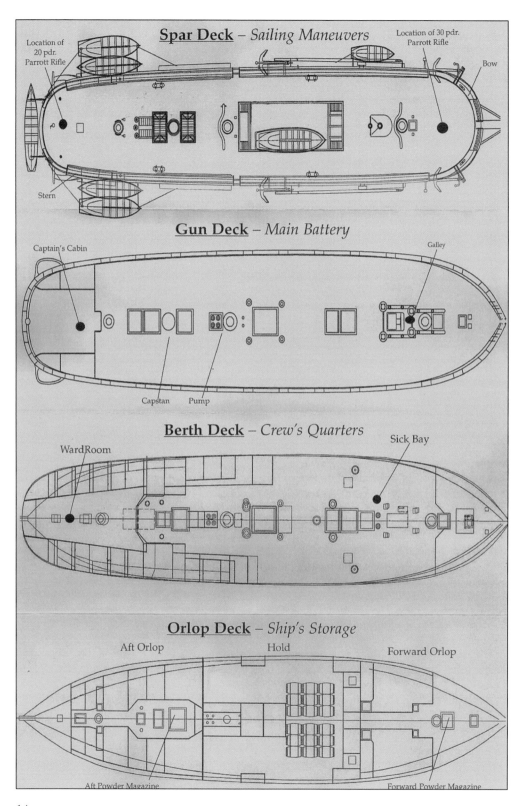

Spar Deck – *Sailing Maneuvers*

Location of 20 pdr. Parrott Rifle

Location of 30 pdr. Parrott Rifle

Bow

Stern

Gun Deck – *Main Battery*

Captain's Cabin

Galley

Capstan   Pump

Berth Deck – *Crew's Quarters*

WardRoom

Sick Bay

Orlop Deck – *Ship's Storage*

Aft Orlop

Hold

Forward Orlop

Aft Powder Magazine

Forward Powder Magazine

14

The uniform of the Marine Corps was constantly undergoing transition. During the 1850s, the uniform was very similar to the army uniform of the Mexican War period. The short jacket was sky blue kersey wool, and the hat was dark blue wool with a steeply pitched leather visor. A script USM insignia centered on the front of the hat was the only distinguishing feature. White buff cross-belts held the cartridge box and bayonet and were secured by the white buff waist belt with a plain, rectangular brass buckle. During her maiden cruise and her service in the African Squadron, the *Constellation's* Marines were armed with .69-caliber smoothbore muskets. During the Civil War, they were armed with Model 1855 rifle-muskets manufactured at the Harpers Ferry Armory. The model here is pictured in a reproduced uniform with the rank of orderly sergeant, the forerunner of the modern day rank of gunnery sergeant. His rank as sergeant is shown by two stripes on the lower sleeve. The addition of the red sash under the waist belt distinguished him as an orderly sergeant.

The first authorization for an official naval uniform came in 1840. Very little changed in the naval uniform other than tailoring changes. During the 1850s, the sennit hat, wide brimmed and constructed of straw, was the issued hat. Many sailors had taken to the habit of lettering the name of their ship on the hat ribbon. The pants were not belled, but cut without a taper to give a wide leg that was easy to roll up for scrubbing or "holystoning" the decks. The frock or jumper was made to be worn outside the trousers, but most sailors preferred to tuck it into their pants. Broadfall trousers with any number of buttons depending on the sailor's whim were commonly worn. The black silk neckerchief was worn under the broad sailor collar of the jumper. Individual taste and sewing ability dictated embroidery, sometimes quite elaborate, that adorned the uniform. No regulation called for striping or stars on sleeves or collar. Therefore, such ornamentation appeared at the choice of the wearer.

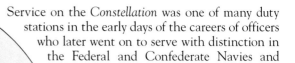

Service on the *Constellation* was one of many duty stations in the early days of the careers of officers who later went on to serve with distinction in the Federal and Confederate Navies and Marine Corps. One of these men was James R. Madison Mullany (left). Mullany was one of the few officers in the Navy who could boast that he had served on both the frigate *Constellation* and the sloop that replaced her. He was ordered to the frigate *Constellation* soon after his appointment as a midshipman in February 1832, and he served with her in the Mediterranean Squadron until November 1834. After rotating through a number of posts, including missing the Mexican War due to assignment to coastal survey duty, Lieutenant Mullany reported to the new sloop *Constellation* in April 1859. He served two months aboard her prior to her departure for the African Squadron.

An unheralded and somewhat forgotten figure, Mullany's service during the Civil War was nothing less than remarkable. He commanded the USS *Wyandotte* in the Gulf of Mexico from November 1860 to June 1861. Promoted to commander in October 1861, he was given command of the 10-gun paddle wheeler USS *Bienville* the following April. The *Bienville* was assigned to the South Atlantic Blockading Squadron. Under Mullany's command, this ship and her crew captured the blockade runners *Providence*, *LaCriolla*, and *Rebecca* south of Charleston.

Mullany reported to the USS *Oneida* on temporary assignment on August 4, 1864, and commanded that vessel in the Battle of Mobile Bay two days later. While fighting the ironclad ram CSS *Tennessee*, Commander Mullany was seriously wounded, resulting in the loss of his right arm. The *Oneida* was heavily damaged in the battle, losing 8 men killed and 30 wounded. With her steering system severely damaged and her boiler exploded, she was towed to safety by the USS *Galena*. Mullany was lauded in the official reports of the battle for his actions. His service was so highly valued by Admiral Farragut that he reassigned Mullany back again to command the *Bienville* a mere two weeks later, a post that Mullany held through the end of the war.

Commander Mullany continued to serve the Navy through the demobilization after the war. From May 1865 to May 1868, he served as the inspector of ordnance at the New York Navy Yard, during which time he was promoted to captain. After serving briefly in several administrative posts, he was given command of the USS *Richmond*, then serving in the European Squadron, which he commanded for 35 months until the end of 1871. While at this post, he received promotion to commodore. Returning home, he spent over two years as the commander of the Philadelphia Navy Yard. He was then promoted to rear admiral and given command of the North Atlantic Station from the USS *Worcester*. Admiral Mullany's final assignment was to the post of governor of the Naval Asylum in Philadelphia. He retired in 1879 with 47 years of service.

In June 1859, the *Constellation*, with Capt. John S. Nicholas commanding, set sail from Charlestown Navy Yard (pictured above in 1875) with orders to join the joint Anglo-American naval force operating on the west coast of Africa. She spent over two years on this station, interdicting ships attempting to run slaves to the Western Hemisphere. A member of a prominent Virginia family and a resident of Owings Mills, MD, Nicholas was appointed midshipman in 1815 and promoted to lieutenant in 1826, commander in 1842, and captain in 1855. His command of the *Constellation* was interrupted when he was sent home on medical leave in June 1861. Other than one brief stint in an administrative post for the Navy Department, he took no active part in the Civil War and died on July 18, 1865. Indicative of the divided nature of Maryland during the war, his son, Wilson C. Nicholas, served in the Confederate Army, rising to captain on the staff of Brig. Gen. Bradley Johnson. Both Captains Nicholas rest in the St. Thomas Church Cemetery in Owings Mills. Unfortunately, no image of the elder Captain Nicholas is known to exist.

The *Constellation* joined the African Squadron in July. In addition to the Royal Navy vessels serving in this force, the American Navy's presence on this station consisted of the 22-gun sailing sloops *Constellation* and *Portsmouth*, the 18-gun sailing sloop *Saratoga*, the 13-gun steam sloop *San Jacinto*, the 6-gun steam sloop *Mohican*, the 5-gun steamer *Mystic*, the 5-gun steamer *Sumter*, and the 2-gun store ship *Relief*. The American force was commanded by Flag Officer William "Jack" Inman (left). Inman designated the *Constellation* his flagship, and commanded the squadron from his quarters aboard her. During the time that he commanded the squadron, 3,600 captives were liberated. A native of Utica, NY, Inman entered the Navy as a midshipman in 1812, was promoted to captain in 1850, and retired in 1867 as a commodore.

Since the *Constellation* served as the flagship of the squadron, the squadron's senior Marine officer was also quartered aboard the ship. Capt. Isaac T. Doughty (right), a native of Poughkeepsie, NY, received his Marine Corps commission at the age of 36 in 1837. He served aboard the USS *Cumberland* in the Gulf of Mexico during the Mexican War, and was almost lost in a freak accident when traveling in a small boat to report for duty on a board of inquiry. He was promoted to first lieutenant in 1847 and captain in 1856. Doughty was the senior Marine in the African Squadron until transferred to the Marine Barracks, Brooklyn, NY, in January, 1860. During the Civil War, he successively commanded the Marine Guard of the USS *Wabash*, led a Marine landing party in the Battles of Forts Hattaras and Clark, and led a force that occupied St. Augustine, FL. He was promoted to major and assigned to command the Marine Barracks, Brooklyn. He was then transferred to command the Marine Barracks, Philadelphia, until his retirement at the close of the War. Major Doughty retired to Poughkeepsie, where he died in 1890.

Landsman William French, pictured here in 1924, a native of Springfield, MA, served aboard *Constellation* during her duty on the African Station from 1859 to 1861. He was discharged from the Navy at the conclusion of his enlistment. During the Civil War, he served in the 24th Massachusetts Infantry Regiment. He served in the post-war occupation of Richmond, marrying a native of Harpers Ferry, WV, whom he met while stationed in the former Confederate capital. The Frenches lived in Harpers Ferry for a few years, then moved to Springfield for the remainder of their lives.

According to French, life on the African Station was miserable. The equatorial climate on the Atlantic coast of the Congo was brutally hot and humid. Liberty was generally limited to two days per six months on St. Helena's Island—Napoleon's home in exile.

Since they were accustomed to working in such an environment, refugees residing in Liberia were hired to do the heaviest work on the ships. Such work included manning the supply boats that traveled between the ships and a supply base. During one period of the ship's service, Inman was ill and confined to quarters on St. Helena's Island. The *Constellation* remained at St. Helena's for six weeks, during which time French was able to get to know "every rock on the island." He managed to come away from the experience with several souvenirs, including a geranium that was growing on Napoleon's former grave and a dogwood stick that he cut from a tree that "grew not 20 feet from his (Napoleon's) hut." In his off hours, French made it into a cane with figures of a man and a woman carved into it. Proud of his handiwork and as a sentimental reminder of his service in the Navy, French kept that cane for the rest of his life.

In addition to the climate, life was quite dull. Although the slave trade was busy, the *Constellation* captured only three ships during her tour. The light ships used to smuggle slaves from African ports were too swift for the large engine-less *Constellation*. Also, the smugglers adopted elaborate communication systems to avoid running into the fleet and were adept at avoiding detection. Nonetheless, French was discharged with $665 in bounties shared from the three captures—quite a tidy sum for the time.

Napoleon's Tomb.

Dr. John Mills Browne was the *Constellation's* surgeon during her service on the African Squadron. He wrote the following letter home to his sister from St. Helena's Island while on liberty:

U.S. Flag Ship "Constellation," St. Helena, Jan. 12th, 1861

My dear Emma: We arrived at this famed oceanic rock nearby a fortnight ago for the object of granting "liberty" to the crew and general recreation and recuperation for the officers, necessarily in a degree of debilitation from African cruising. That the visitation at this port is enjoyed needs hardly require mention. Nice the mere fact of being at a civilized port. Learning our language spoken & the absence of male Negroes are features in themselves (to us) attractive. If to these be added the interest attached to the name of Napoleon, the amount of present gratification can be imagined.

I have visited "Longwood," the "tomb," "Hutt's Gate," Plantation Houses & the Briers. Situations conversant to the reader of Bonaparte's life & there fore need not share description within the limits of this letter. The details will be reserved for our not far distant meeting. I will only mention that the tomb, its valley and "Longwood" are now the property of the present Emperor. They have had extensive repairs effected & are under the guardianship of a French Colonel, assisted by a Sergeant and a Corporal. Enclosed you will find floral souvenirs of my visit - said visit being incorporated with a picnic given by our mess in the valley of the tomb as seen in the sketch (shown above). This island possesses scenery unexcelled in grandness and loveliness. I have devoted my days on horse to its inspection & am topographically informed of the greater portion of the island. We are receiving the hospitalities of the English Army officers & others of the populace. In about a week, we will probably leave for Loandra & have another short cruise off the Congo. My kind remembrances to Mr. Holton & Ugrid & please remind the children now & then of their African male. I am my dear Emma, affectionately yours...

To encourage productivity among the crews, the Navy Department authorized a bounty of $25 for each slave liberated. During her service off the African Coast, the *Constellation* captured three ships, the brigs *Triton* and *Delicia* and the barque *Cora*. The first two, although outfitted for slave running, had no cargo aboard. The capture of the *Cora*, however, liberated over 700 captives, who were resettled in Liberia. The *Cora* was spotted leaving port and heading to sea on September 26, 1860. Inman ordered that she be chased. The odds of the big warship catching up to the smaller, faster vessel were poor.

According to Landsman French, "Commodore Inman called on the entire crew to trim the vessel for the chase. He got up the carpenter to spruce up the rigging and several of the crew manned pumps to wet the sails so they would push the old sloop along. Once in a while we'd fire a shot, but that didn't scare them any, because we didn't try to hit them. Then I heard Commodore Inman telling...what would happen after darkness fell. 'I know what he'll probably do,' he said. 'He'll square his yards and think he's leaving us on this course. But we'll just fool him by squaring our own yards the minute it gets dark, and keep right up with him.' The minute it became dark, Commodore Inman ordered the course changed, and we nearly ran the *Cora* down. Our jib was almost in her rigging when the lookout shouted a warning."

A lieutenant was placed in charge of a prize crew, which opened the barque's hatches to discover 705 frightened, cringing souls. None of the crew would confess to being the captain of the vessel, but it was soon determined that the ship's owner was aboard. French related "it was a fearful job, cleaning and doctoring those natives. They were nearly starved, but they responded to treatment and, after keeping them a while, we landed them in Monrovia, Liberia, where according to custom, they were colonized." The captured barque was sent on to New York where it was auctioned, the crew released and the captain placed under bond. Slaves liberated by ships of the Royal Navy were transported to the West Indies and indentured to pay for the cost of maintaining the British anti-slavery squadron. The U.S. Navy did not engage in this practice.

On May 21, 1861, a month after the bombardment of Fort Sumter, the *Constellation* seized the *Triton*, which was registered in Charleston, SC. No slaves were aboard. Therefore, unwittingly the *Constellation* affected the first capture of a Confederate ship, naval or merchant, during the war.

Lt. Alexander Colden Rhind was another of the *Constellation's* officers who went on to perform meritoriously during the Civil War. Rhind served aboard the *Constellation* for two and a half years, during her service in the African Squadron, from April 1859 to October 1861. In 1862, he commanded the USS *Crusader* when it supported the noted African-American ship pilot Robert Smalls who, commanding the *Planter,* attacked Fort Johnson in June. Fort Johnson was an earthen battery located on James Island, SC, south of Charleston. In August, Rhind was transferred to the USS *Seneca,* and later that year was ordered to the ironclad USS *Keokuk* as her commanding officer. On April 7, 1863, the *Keokuk* participated in an attack on Charleston, SC, by a flotilla of nine monitors.

The *Keokuk* led the advance and was within 550 yards of Fort Sumter before the attack was broken up by heavy fire from Confederate shore batteries located on Morris Island, Sullivan's Island, and from Fort Sumter itself. Rhind was slightly wounded in the attack, and the *Keokuk* was so heavily damaged that it sank the next day.

Rhind was then sent to the USS *Paul Jones,* and he commanded that ship until ordered to the USS *Wabash* in September; he later was given command of the USS *Agawam* in the North Atlantic Blockading Squadron, holding that post into 1864. On December 23, 1864, he undertook a dangerous mission of commanding the powder boat USS *Louisiana*. The goal was to bring the *Louisiana,* laden with gunpowder, as close to Fort Fisher as possible, abandon the ship, and detonate her stores in hope of destroying the fort in the blast. The ship was destroyed, but the fort was not. For his leadership in this mission, he received a commendation. Commander Rhind later participated in the Neuse Expedition of 1865 in support of the U.S. Army forces at North Edisto, CA. Rhind retired from the Navy in 1883 with the rank of rear admiral.

In January 1860, Marine Captain Doughty was replaced by First Lt. John Rodgers Fenwick Tattnall (left), who reported from the USS *Portsmouth*. A native of Georgia, Tattnall served on the *Constellation* until being transferred to the USS *San Jacinto* on August 1, 1861. Unaware that Georgia had already seceded, he wrote to Governor Joseph Brown from the *Constellation* on February 28, 1861, offering his services to that state, should it withdraw. Once aboard the *San Jacinto* in August, Tattnall attempted to resign. That ship's commander, Capt. Charles Wilkes, demanded Tattnall's sword. Rather than surrender it, Tattnall threw it overboard. He was subsequently arrested. Exchanged as a prisoner of war in 1862, Tattnall served briefly as colonel of the 29th Alabama Infantry. Proclaiming "I would rather command a company of Marines than a brigade of volunteers," he resigned to become the captain of Company E, of the Confederate Marine Corps. He served in that capacity through the end of the war. A post-war resident of Savannah, GA, he died in 1907.

The *Portsmouth* was also a 22-gun sailing sloop. Here, she is pictured during the Civil War with the French Navy's steamer *Milano*. After her service in the African Squadron, she served most of the Civil War in the West Gulf Blockading Squadron under Admiral Farragut and participated in the capture of New Orleans. She served into the 20th century, when she was stricken from the rolls. As a means of disposal, she was intentionally set afire and burned by the Navy in 1915.

Liberia was established in 1822 as a haven for former slaves who desired to return to Africa. Its capital of Monrovia (shown here) was named for American President James Monroe. When ships of the African Squadron captured a slaver and liberated its captives, the U.S. Navy transported the victims to Monrovia for colonization. Although most of the captives were not native to Liberia, this was the only viable solution. Since the slaves sold and transported out of the Congo River Valley were from all over central Africa, it was impossible for the Navy to determine the precise origin of the overwhelming majority of former captives due to language barriers and other factors.

When Captain Nicholas was sent home on medical leave, Capt. Thomas A. Dornin (left) transferred to command the *Constellation* in June 1861. Soon thereafter, the entire squadron, except the USS *Saratoga*, was ordered home due to the outbreak of the rebellion. The *Constellation* dropped anchor in Portsmouth, NH, on September 28, 1861. Although the anguish of the brewing war at home undoubtedly plagued the minds of all hands, no major problems occurred aboard her during this time. The overwhelming majority of the crew were New Englanders. Therefore, the crew was most assuredly in support of the Union.

A native of Ireland, Dornin entered the Navy as a midshipman in 1815, rising through the ranks to captain in 1855. While in command of the USS *Portsmouth* in 1853, he and his ship were charged with preventing the adventurers under William Walker in Nicaragua from invading Mexican territory, as Walker was attempting to seize Baja California and Sonoma. Dornin later commanded the Norfolk Navy Yard from 1855 to 1859. In 1862, he was made commodore, commanding the navy yard at Baltimore for most of the war.

The 13-gun steam sloop *San Jacinto*, commanded by Capt. Charles Wilkes, also served in the African Squadron prior to the Civil War. While returning to the U.S. pursuant to orders, Wilkes' actions sparked an international incident that almost led to war with Great Britain. On November 8, 1861, while in international waters, the *San Jacinto* stopped the British mail steamer *Trent* and forcibly removed James M. Mason, John Slidell, and their two assistants to Wilkes' ship. Mason and Slidell were the Confederate government's emissaries to Great Britain and France. Popularly known as "The *Trent* Affair," this action embarrassed the Lincoln administration. France and England took umbrage at this violation of maritime law and British war hawks urged war with the United States. Facing the possibility of having to fight England as well as the Confederacy, the State Department apologized and released the prisoners to British officials.

Virginia native Donald McNeill Fairfax (left) served in the African Squadron aboard the *Mystic* and as the *Constellation's* executive officer (second in command) before transferring to the *San Jacinto*. When Wilkes stopped the *Trent*, Fairfax was sent aboard to determine whether anyone tied to the rebellion was on board. Finding Mason and Slidell and their entourage, he reported this information to Wilkes, who ordered the men seized. Fairfax, seemingly embarrassed by the whole affair, executed his orders and escorted the party onto the *San Jacinto*. A Virginian who stayed loyal to the Union, Fairfax commanded a succession of ships in 1862 and 1863, including the monitor USS *Montauk*. He served as commandant of midshipmen at the Naval Academy from September 1863 to October 1865, during its brief relocation to Newport, RI. The son-in-law of Admiral Andrew H. Foote, Fairfax died at his home in Hagerstown, MD, in 1894.

When the *Constellation* arrived at the Portsmouth Navy Yard at Kittery, ME, those crew members whose enlistments had expired were discharged from the service. Also, the officers and crew received their share of the bounty from the capture of the slave ships. The *Constellation's* commissioned officers were reassigned and new staff arrived to prepare her for her next assignment.

The tiny dramas occurring in the officers' quarters at almost every fort, army post, and marine barracks as well as aboard the ships of the navy also played out in the *Constellation's* ward room. Brother officers bade farewell to one another and chose their cause for the coming Civil War, unaware of the carnage that lay ahead.

Two of the officers who did not continue in the Navy were Lt. Benjamin Loyall and Midshipman Walter Raleigh Butt (above). Loyall and Butt left the Navy to cast their lot with the South and the new Confederate States Navy. Temporarily imprisoned aboard the USS *Congress* and at Fort Warren in Boston, Butt (shown here in his Confederate Navy uniform) was exchanged and then commissioned a lieutenant in the Confederate Navy. He was assigned to the ironclad CSS *Virginia* (formerly the USS *Merrimac*) when she was scuttled to prevent her capture in May 1862. After service overseas, Lieutenant Butt returned to the South and reported to the James River Squadron, which was protecting the capital at Richmond. He was temporarily assigned to command the steamer CSS *Hampton,* and later commanded the gunboat CSS *Nansemond* at the close of the war.

Two

THE COURSE OF THE STARS

Gideon Welles

SECRETARY OF THE NAVY

Although lacking any real nautical experience, Secretary of the Navy Gideon Welles possessed good sense and a fierce loyalty to President Lincoln. This former newspaper editor from Glastonbury, CT, relied on the vast experience of Assistant Navy Secretary Gustavus Fox. It was the order from Welles that sent the *Constellation* to patrol the Mediterranean, thus removing her from the active seat of war. Her mission was to assist in capturing or destroying the CSS *Sumter*, then reported near Spain. Her secondary mission was to protect American commerce while "showing the flag." The Europeans were quick to take advantage of the American Civil War and its apparent weakening of the Navy in overseas protection of merchant vessels. Thus, Welles was forced into planning global strategy as well as domestic war efforts.

The stern countenance shown by Cmdr. Henry Knox Thatcher (1806–1880) matches his sour attitude to being assigned a sailing vessel during an age of modern movement to steam power. Thatcher was a strict disciplinarian whose severe control made him very unpopular with his crew despite their respect for his ability as a fighter. Having a particular dislike for one seaman accused of malingering, Thatcher ordered the man lashed into his hammock and placed next to the galley. The heat soon caused the man to perspire so extensively that his sweat permeated his mattress and hammock and flowed onto the deck. The man had to be discharged for medical reasons a few days later. Yet, Thatcher was an able seaman who knew how important control was in running a man-of-war during a time of crisis. Upon arrival at Algiciras, he discovered that the CSS *Sumter* was rotting away in port under the watchful guard of the USS *Tuscarora*. Making full use of his recent promotion to commodore, Thatcher left the *Tuscarora* to her boring duty, resuming his patrol of the Mediterranean.

The grandson of Maj. Gen. Henry Knox, Washington's artillery chief in the Continental Army and later the first secretary of war, Captain Thatcher was a native of Thomaston, ME. He briefly attended the Military Academy at West Point before accepting an appointment as a midshipman in the Navy in 1823. Over the next 38 years, he served in each of the Navy's squadrons, serving all over the world, earning promotion to commander in 1855 and serving as executive officer of the Boston Navy Yard at the outbreak of the war.

In September 1863, Thatcher was ordered to the USS *Colorado* to take command of the 1st Division of the North Atlantic Blockading Squadron, serving in that post until February 1864. For the next six months, he served as temporary commander of store ships at Port Royal, SC, before being ordered back to his post on *Colorado*, commanding a division of the squadron in the Battle of Fort Fisher in January 1865. Promoted to acting rear admiral on January 25, 1865, Thatcher was given command of the West Gulf Blockading Squadron for the duration of the war. After the war, he commanded the Gulf Squadron from the USS *Estrella*. Promoted to rear admiral in July 1866, Thatcher's last assignment was command of the North Pacific Squadron from the USS *Vanderbilt* until May 1868.

By the time the *Constellation* reached Spain, the CSS *Sumter* (shown here capturing the merchant vessel SS *Joseph Parke*) had reached the end of her usefulness as a raider. During just seven months at sea she had managed to take 18 vessels. Unable to make repairs and blockaded in harbor by the U.S. Navy, she served the sole purpose of forcing vessels of the U.S. Navy to stay nearby to assure that she would never again wreak havoc on Northern shipping. She was later sold into Spanish registry and renamed the *Gibraltar*. Despite U.S. protests, the *Gibraltar* was soon back in the service of the Confederacy in mid-1863 and continued until she was laid up in England for repairs in 1864. She foundered near Cherbourg, France, in 1865 and lies very near the CSS *Alabama*.

Commanded by Capt. (later Rear Adm.) Raphael Semmes of Charles County, MD, the *Sumter* was one of the major early success stories of the fledgling Confederate States Navy. Semmes, a veteran officer of the old U.S. Navy, including a stint on the frigate *Constellation* in the 1830s, would be heard from again as the captain of the fabled Confederate raider CSS *Alabama*. When the Confederate capital at Richmond was evacuated in the waning days of the war, all Naval and Marine Corps personnel were ordered under Semmes' command to march west with Robert E. Lee's Army of Northern Virginia. For this reason, Semmes was commissioned brigadier general in the army, making him the only man in the Confederate service to hold both the position of admiral in the navy and general in the army.

The USS *Tuscarora* was the principal U.S. vessel keeping watch over the CSS *Sumter*. The skeleton crew aboard the Confederate raider enjoyed taunting U.S. vessels guarding her, as the crew of the *Constellation* soon learned upon arrival. To overcome the boredom and monotony of guard duty, the *Tuscarora's* crew was allowed to stage theatricals and musical productions. The crew of the *Constellation* were allowed to attend one of these evening productions. Just before the third act, Lt. Cmdr. Lowe, executive officer of the *Constellation*, ordered his crew to return to ship despite there being no emergency need. This inspired diarist Moses Safford to record, "Such is life aboard our ship." As in the case with Commodore Thatcher, there was little regard among the crew for Commander Lowe as well. It was customary for the crew to cheer the captain of the ship as he departed for a new assignment. When Thatcher and Lowe left the *Constellation* in July 1863, their departure was observed by the crew with an icy silence.

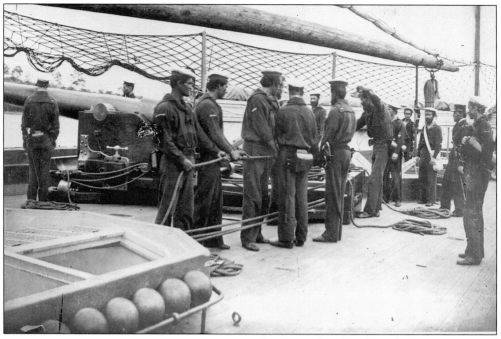

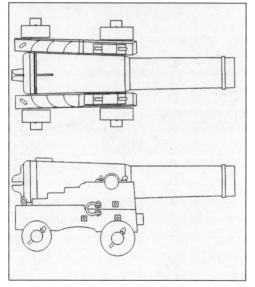

Supplementing her broadside armament of sixteen 8-inch shell guns and four 32-pounders to deal with Confederate commerce raiders, was a 30-pound Parrott Rifle mounted on the forward spar deck and a 20-pound Parrott Rifle mounted on the aft spar deck. No photographs of the *Constellation* are known to exist from the Civil War period. This photograph of sailors drilling on a Parrott Rifle on the spar deck of the USS *Mendota* is very similar to what would have been seen on the *Constellation* at this time. Here, the gun is trained to starboard with the spanker boom over the crew's heads. Note the metal rails laid in the deck to guide or "pivot" the heavy gun's wheels from one position to another.

Sixteen of the twenty guns mounted on the gun deck were 8-inch "shell guns," chambered to fire 32-pound ammunition. These guns were the more traditional-style "broadside guns" for which the battles of the old sailing navies were famed. While the guns on the spar deck were used more for isolated shots and when the most accuracy possible was necessary, these guns were the mainstay battery of the ship. They would have been used to concentrate fire on an enemy ship or position. There was only a sufficient number of crew to serve one side of the ship's battery at a time. Seldom, if ever, would a ship need to fire both sides in an action and, if so, the gun crews would divide up to service the guns on the opposite side. Combatants maneuvered their vessels to fire broadsides into the weakest part of a sailing warship; the bow and the stern. Here few guns were mounted to fire at an attacking enemy and a raking fire could run the length of an enemy ship's deck causing the most damage possible. The remaining four guns on this deck were standard 32-pounders.

ARTILLERY PRACTICE WITH THE DAHLGREN HOWITZER BOAT GUN—OFFICER GIVING THE WORD OF COMMAND TO FIRE.

In addition to the permanently affixed guns, the *Constellation* was armed with three Dahlgren 12-pounder boat howitzers. These guns were kept dismantled, with their iron carriages, in the ship's boats. They were also designed to be used ashore. Three models of this gun were produced. The *Constellation's* battery of howitzers consisted of two of the "heavy" model, weighing in at 750 pounds each, and one "light" model, at 420 pounds. None of the smaller 320-pound model were assigned to the ship. Here, an artist depicts a group of naval personnel training on the howitzer.

For organizational purposes, the crew was divided into several "divisions." Each division and the Marine Guard were regularly drilled in the use of the broadside guns—the Parrott Rifles and the howitzers—as well as rifles, handguns, and (except for the Marines, who perfected the bayonet drill) cutlasses and pistols. The ship's logs (above) are in the possession of the National Archives. These logs detail the mundane day-to-day official happenings on the ship, including notations of when the Marine Guard and the various divisions were drilled as well as the types of drill in which they participated.

Moses Safford served as ship's yeoman from 1862 to 1865. As yeoman, he was essentially the chief noncommissioned officer in charge of the ship's supplies. Much that is known about the *Constellation* comes from an elaborately detailed diary that he kept during his service. His position as a staff NCO in charge of a busy post, who worked closely with the crew, put him in a unique position to be an excellent observer of all the ship's happenings. Safford's account reached across a century in solving a mystery. During the planning stage for the restoration of the ship in the early 1990s, it was unknown whether the *Constellation's* portholes were installed during her initial construction or after the Civil War. The debate was ended with the discovery of the following notation in Safford's journal dated September 13, 1863:

> *Last night the wind increased to a strong gale... Some of those trying to sleep on the berth deck last night received surprises unpleasant to them but amusing to others. The berth deck air ports were open as is customary in port. As it became rough and the ship began to tumble about Boatswain Hunter was awakened by a stream of water the full size of the air port which came in and struck him in the chest and face. A wave dashing through the opening in the Dispensary doused the Doctor's Steward who was sleeping there, soaked his desk and papers and made it necessary for Smith to give up sleep and start bailing out the room. A wave came through another port and over the 'bag rack' and struck 'Boy' Hurley and filled his hammock with water.*

Safford signed on for the cruise as the ship's yeoman at the personal behest of Captain Thatcher. Safford's writings hint at frustration or disappointment that, as a man of education and legal training, he was relegated to a noncommissioned officer's rank. His journals detail the petty wranglings among the ship's officers, the fisticuffs among the crew, and many funny, unique, and poignant happenings that occurred on the ship during its 33 months of duty during the war.

Safford was discharged in February 1865 and resumed his law practice. Later he became a clerk in the engineering section of the Portsmouth, NH, Navy Yard during the McKinley Administration.

It was not uncommon for a man to spend his entire working life as a simple sailor in the Navy or merchant fleet. There was little room for advancement; therefore, positions of prestige were determined more by duty assignment on a ship than through promotion. The men who worked aloft, particularly the topmen, were the top dogs of the ship's community. Next came the forecastlemen, then the afterguard, and finally, the waisters. Largely, position was dictated by service and age. Oddly, this arrangement was contrary to the common belief that age and experience are preferred. The younger, more agile sailors who had not yet matured to a point to realize that they were indeed mortal, worked in the tops. Old salts, such as Moses Webber (pictured here), were more likely assigned to a gun crew in the middle or "waist" of the ship.

Although African Americans participated in the Navy even before the birth of the United States, the Civil War is when they received their first official recognition. From the very beginning of the war, shortages in manpower were recognized, and the Navy Department issued a formal declaration that black sailors would be welcome and openly recruited for naval service. A formula dictating one black sailor for every 20 whites was widely ignored. The *Constellation* had a minimum of 10 to 15 black sailors aboard and most likely there were several more. At least three of these men died on board while serving their country during the Civil War years. Although this sailor is definitely not from the *Constellation*, he makes a good representation of all African Americans who served. By 1863, with the issue of the Emancipation Proclamation by President Abraham Lincoln, black sailors had an even greater cause to guide their noble service. When the *Constellation* became a receiving ship in 1865, her crew was over 70 percent African American.

Fortunately for the U.S. Marines, the Navy prevailed in having the Corps maintained. The War Department had considered disbanding the Marine Corps as a fiscal savings effort, but the Navy insisted that the Marines were necessary for maintenance of order aboard ship as well as useful in amphibious operations. The Marine Guard on the *Constellation* consisted of one commissioned officer in command of 2 sergeants, 3 corporals, 36 privates and 2 musicians. Unlike the integrated Navy and the segregated Army, African Americans were not taken into the Marine Corps at this time.

The uniform regulations changed considerably in 1859. The standard daily wear undress coat became a frock coat of dark blue cloth with a red welt around the base of the collar and seven buttons down the front. The cap was changed to the chasseur-style kepi also of dark blue with a brass hunting horn device bearing a red center with an Old English letter "M" superimposed. Plain sky blue trousers were worn in winter and white cotton trousers were the standard wear in summer and when in tropical areas. Chevrons consisted of gold cloth taping on a red background, worn "points up," in contrast to the Army, which wore them "points down." A pullover dark blue fatigue blouse was also authorized for work details. The dress coat was quite fancy, with gold lace tape and red trim on the collar and cuffs. A shako bearing a shield device finished the appearance.

Marines aboard the *Constellation* were armed with the Model 1855 rifle-musket, manufactured at the armory in Harpers Ferry, VA (now WV). Accouterments to serve the weapon were made of black leather and carried on a waist belt and crossed belts of white buff leather. Officers wore a similarly plain frock coat adorned with Russian knots on the shoulders bearing the same insignia for rank as then in use for the Army. The officer's kepi wasdecorated with a red cloth insignia bearing a gold embroidered hunting horn device that also encircled the letter "M." By regulation, officers carried their swords on white leather sword belts. The models in this photograph display the summer undress uniform of a lieutenant and sergeant of Marines.

With so much time spent on various port calls during her service in the Mediterranean area, many of the officers and men of the *Constellation* sat for painted and photographic portraits while in port. Yeoman Safford related that, when it became known that the ship would soon leave a port, officers and men alike would go on buying binges for underpriced goods and services. Since most of the crew hailed from New England, who knows how many mementos and images lay forgotten in many a Massachusetts or New Hampshire attic or trunk? On the following pages, you will meet several of the officers and senior noncommissioned officers who served aboard the *Constellation* during her duty in the Mediterranean during the Civil War.

Edwin H. Miller (left) began the cruise as a quarter gunner, was subsequently promoted to master's mate in early 1863, and later to lieutenant during his service aboard the ship. Here, he is attired in his ensign's or "master's" coat. A master's mate was a position of a warrant officer. Master's mates were appointed to this post only after serving a minimum of one year of successful sea duty and at the recommendation of their superior officers. Miller managed, through exemplary performance of his duties, to be promoted to a position as a line officer.

A native of Massachusetts, Midshipman Charles F. Blake (right) left the Naval Academy (temporarily located in Newport, RI, for the duration of the war) to report for service aboard the *Constellation* in November 1861. He served aboard the ship from the outset of the Mediterranean cruise until transferred back to the U.S. in May 1863, in order to take the examination for master. The proof of the Navy's severe need for qualified officers is revealed by Blake being removed in the midst of active service to take the necessary written and oral examinations for advancement.

When the *Constellation* put to sea in March 1862, her Marine Guard was commanded by 23-year old Second Lt. Robert O'Neil Ford. A Philadelphian by birth and a resident of Brooklyn, NY, Ford received his commission a mere 10 weeks before reporting to the ship in November 1861. He served briefly in the elite 7th New York State Militia Regiment before receiving his Marine commission on August 30, 1861. He reported to Marine Corps Headquarters, serving there until ordered to Kittery, ME, to join the ship.

Ford's role aboard the ship was limited. The orderly sergeant handled the day-to-day operations of the 42-man Marine Guard. Ford's role was confined to presenting the executive officer with the daily report of the Marine Guard and commanding the Marines in drill, ceremony, and combat.

Yeoman Safford seemed not to care for the young Marine. On February 29, 1864, upon entering the harbor at Spezia, Safford wrote: "The Lieutenant of Marines was, as usual, on deck dressed for the shore as soon as land was distinctly made out. Until then, he had not been seen on the deck during the passage. He is always ready to step on terra firma and is of quite as much service to the ship there as anywhere else." Nonetheless, he was promoted to first lieutenant in July 1863.

Removed as he was from the daily operations of the Marine Guard, Safford's diary reveals only the extraordinary events associated with the leathernecks. He mentions in his observations a time when Ford mistook the crew's pet hedgehog (which had gotten loose in the wardroom) for a rat, and attacked the creature with his sword. It is apparent from Safford's journal that Ford did not get along well with several of his fellow officers, including the purser and the ship's surgeon, Dr. Messersmith. On September 1, 1864, Ford was engaged in a drunken brawl with Messersmith, Acting Master Kempton, and the ship's paymaster in the streets of Spezia.

Upon the *Constellation's* reassignment to Norfolk, Ford was transferred to the Marine Barracks at Brooklyn. He resigned his commission in 1868 and resided in Brooklyn until his death in 1913.

As the *Constellation* navigated the Mediterranean, traveling between Spain, Italy, Greece, Lebanon, Palestine, Egypt, and Tunisia, there was much opportunity for sightseeing while in port. Specialty positions on board qualified many men to serve as staff officers who were given more free time in port than most. In company with Yeoman Safford, Purser's Steward William E. Cox (above left), who later was promoted to paymaster's clerk, went on numerous excursions much as modern-day sailors and marines continue to do today while on various duty stations. At a time when Americans were dying by the thousands at home, these men wrote of the tedium and boredom of their jobs and broke the monotony by taking tourist trips.

Boatswain John R. Hunter (below left) appears in this portrait in his formal uniform. His rank is indicated by the narrow strips of gold lace that contain an old English letter "B." Only a star adorns the lower part of each sleeve of his frock coat, and he is wearing his summer white tropical trousers. The sword is most likely a photographer's prop. Capitalizing on the background of Yeoman Safford as a competent seaman and accepting Safford's offer to serve as a teacher, former Boatswain's Mate Hunter ably received promotion to boatswain.

The *Constellation's* guns saw most of their use in ceremonial occasions as numerous VIP guests were either welcomed aboard or were saluted in port. Despite their lack of use in firing shots of anger, the gunners kept the guns spotless and in good repair. Once cleaned, they were painted and varnished to make them shine. For example, Quarter Gunner Charles "Cocky" Anderson was notorious for keeping his section of guns spotless and taking vengeance upon anyone who would dare touch them without his permission. Less finicky, but just as proud of his assigned ordnance, was Gunner John R. Grainger, pictured here.

Grainger was the senior gunner on the ship. He was widely respected among the officers and crew for his prowess with the use of the 30-pound Parrott Rifle on the spar deck, with which he was a crack shot. His expertise and judgement was so valued by the ship's officers that when the ship was in port and gunpowder had to be purchased for the ship's guns, Grainger was ordered to go with the officers to inspect and approve the goods prior to their purchase.

Emmet Barnes was a boatswain's mate who also took in the sights. He is pictured here in civilian attire. He served as a storeroom clerk aboard ship, a position that was labelled "jack of the dust." While on liberty (predominantly in Italy), many of the ship's company purchased souvenirs, personal goods, and civilian clothing items, thus taking advantage of both European quality and cheaper prices. Many sailors often lamented their service and constantly spoke of how they longed to be civilians again. Yet after sampling the realities of civilian life even for a short period of time, many simply rejoined the Navy, where they felt more at home once back at sea.

Duty assignments would often change for naval officers. Many times orders would arrive via packet boat bringing news that some officers were to be transferred stateside to fill vacancies created by the casualties of war. Changes in uniform regulations, which occurred officially on at least three separate occasions during the war period, were not followed too closely. Consequently, numerous photographs exist of naval officers in various combinations of early and late war regulation dress. Acting Master Eugene B. Mallett, shown here, is an excellent example. He is wearing the late war coat with its two rows of 1/4-inch sleeve braid and yet his cap is that issued prior to the war with its horizontal anchor device and a wide band of gold braid around the crown.

When Commodore Thatcher was ordered to return to the United States, his vacancy on the ship was filled by Capt. Henry S. Stellwagen (1810–1866). A native of Philadelphia, Stellwagen (opposite page) entered the Navy as a midshipman in 1828. During his naval career, he became heavily involved in the survey of the Atlantic coastline and is credited with the discovery of a subaquatic feature north of the tip of Cape Cod that was later named for him. "Stellwagen Bank" is now a National Marine Sanctuary. He was also recognized and received an award from the Franklin Institute for inventing a device, the "Stellwagen Cup," that was used to retrieve sediment samples from the ocean floor.

A veteran of Commodore Perry's fleet during the Mexican War, Stellwagen's assignment to the *Constellation* may have been, to some degree, a punishment. In 1862, Stellwagen commanded the steamer USS *Mercedita* in the Gulf Squadron and is credited with capturing seven blockade runners. He and the ship were transferred to the South Atlantic Blockading Squadron and assigned to the blockade of Charleston Harbor. At 4:30 a.m. on January 31, 1863, the *Mercedita* was caught by surprise and rammed by the CSS *Palmetto State*. Crippled from this action, she was only able to offer token resistance. With her hull pierced and her port boiler ruptured, Stellwagen struck his colors. The captain of the *Palmetto State*, believing the *Mercedita* to be doomed, accepted paroles on behalf of the officers and crew and steamed on to other nearby targets. The crew effected emergency repairs and the *Mercedita* limped into Port Royal, SC. She was later repaired in Philadelphia and placed back into service in April.

Stellwagen was fairly popular with the crew, especially in the wake of Commodore Thatcher's management style. In the company of his son Thomas, whom he employed as his clerk, Stellwagen engaged in the same sightseeing as much of the crew. Yet, he proved his eagerness for action by promptly dispatching his ship in pursuit of the reported Confederate cruiser *Southerner*.

No single ship served the entire duration of the war on the European station as numerous ships rotated in and out of that region. Despite missing real combat action during the war, the *Constellation* worked in close cooperation with several ships that did go on to glory. In April of 1862, the USS *Kearsarge* (shown above, c. 1864) was stationed with the *Constellation* near Spain. Capt. Charles Pickering, who was later replaced by Capt. John Winslow, even made a visit aboard when the ships were together at Cadiz. The *Constellation* had received orders to pursue and destroy a new commerce raider known as the *Southerner*. Naval records indicate that the *Southerner* was a wooden steam ship built on an iron frame, sailing under British registry. Despite several active searches, the *Southerner* was never located.

Later, in 1864 while the *Constellation* was in Palestine, the *Kearsarge* engaged in the famous battle near Cherbourg, France, which led to the sinking of the CSS *Alabama*. On board during the combat was former *Constellation* surgeon Dr. John Mills Browne. Fortunately, Browne had few casualties to attend as only three men from the *Kearsarge* were injured during the fight. However, he did have to see to the welfare of the Confederate crewmen captured subsequent to the action. Such mercy shown to former enemies was common in this war of Americans versus Americans.

Strangely, the *Constellation* was inadvertently called upon to serve a mission of peace during a time of war. Upon being summoned back to the States in 1864, the ship came across the brig *Mercey*, which had been devastated in a hurricane. The desperate crew was barely able to handle their ship when the *Constellation* hove into sight. The warship made every effort to see to the brig's welfare by caring for her injured crew and giving a needed tow. Only after the *Mercey* was safely in port at St. Thomas did the *Constellation* return to her orders to report to Admiral Farragut at Mobile Bay. In gratitude, the British government later presented Stellwagen with a commemorative sword.

On November 27, 1864, the *Constellation* arrived on station at Mobile. Upon reporting to Admiral Farragut (below), Captain Stellwagen was informed that his ship's deep draft was too much for her to be of service here. She was ordered then to Norfolk. During a brief layover in Pensacola, she exchanged salutes with Farragut's flagship, the USS *Hartford* (pictured at right from an 1877 stereo card) and with the ship that would soon replace her, the USS *Potomac*. Her days of active service numbered, the crew was soon depleted by the transfer of personnel to other active duty ships, particularly to the USS *Kennebec*.

After leaving Pensacola, the *Constellation* sailed for Norfolk, via Havana, Cuba. On the way she pursued an unknown vessel hoping to nail down a prize. The fruitless chase ended with the ship coming into port at Havana, Cuba. Havana being a sanctuary for Confederate vessels, the ship was assailed verbally by enemy crews as she came to her anchorage. Liberty granted the crew resulted in a brawl ashore with Confederate crews. Captain Stellwagen was little amused when his crew returned on board wearing an assortment of captured Confederate clothing. Fully understanding their "trophy collection," the men received little to no punishment and the ship upped anchor to resume her voyage.

Upon leaving Cuba, the ship proceeded up the eastern seaboard en route to Norfolk. Off the coast of North Carolina, a blockade runner was spotted making for the open sea. Too far off to effect a successful chase, Captain Stellwagen ordered shots to be fired in her direction. These shots were the only ones fired in anger by the *Constellation* during the entire war. Firing these shots accomplished the desired result. Other U.S. Navy vessels in the area were alerted by the cannon fire. The blockade runner was captured by the other ships, but since the prey was chased beyond the horizon, the crew of the *Constellation* lost out on any share of the bounty that the captured ship brought.

On Christmas Day, 1864, the *Constellation* entered the Chesapeake Bay and reported to the navy yard at Norfolk, VA. In January, the crew and the ship's boats were employed to ferry sailors and Marines that were wounded in the recent Battle of Fort Fisher, NC, from ships to the naval hospital. Only then did many of the crew realize how fortunate they had been to be assigned to a foreign assignment for the previous three years. Those of her crew and Marine Guard whose enlistments had expired were discharged from the service. Stellwagen and the rest of the officers were soon reassigned. At this point, the *Constellation* began her service as a receiving ship used to house and train newly recruited sailors. She remained at Norfolk until June 1866, when she was transferred to Philadelphia. She was decommissioned late in 1868 and her logs were transferred to the receiving ship *Potomac*.

During part of her time as a receiving ship at Philadelphia, the *Constellation*'s Marine Guard was commanded by First Lt. Henry Clay Cochrane of Chester, PA. He had already seen much action since receiving his commission in 1863. Initially assigned to Marine Corps Headquarters, one of his first duties was to escort the Marine Corps Band to Gettysburg to participate in the dedication of the National Cemetery. Traveling by train, President Lincoln escaped his entourage and visited with the Marine officers in the car to which the Marines were assigned, sharing a seat with Cochrane until they arrived in Baltimore. Cochrane marched behind the President in the procession to the cemetery and was on the rostrum as Lincoln gave his immortal "Gettysburg Address."

Cochrane later served at the Mound City Marine Barracks in Illinois, and then commanded the Marine Guard of the USS *Black Hawk* in the Mississippi Squadron from March to May, 1865. When the *Black Hawk* caught fire as a result of a coal handling accident, he and his Marines were credited with effecting a safe evacuation of the ship's crew, waiting until the last minute to jump overboard. After recruiting duty in Chicago, Cochrane was transferred to the Philadelphia Navy Yard in 1866. He was reassigned within the Philadelphia post to the Marine Guard of the receiving ship *Constellation* from October 1867 to July 1868.

THE FRIGATE NEW IRONSIDES.

One notable event occurred during his assignment to the Philadelphia Navy Yard and subsequent assignment to the *Constellation*. On the night of December 10, 1866, the seagoing ironclad frigate *New Ironsides* caught fire in the Delaware River. Due to the emergency, Lt. Cochrane led a party of Marines gathered from around the Philadelphia Navy Yard, including Marines from the *Constellation*, and a party of firefighters on a 6-mile nighttime march to the site of the disaster in an attempt to save the ship from destruction. The party boarded the ship twice, combating the flames, but the fire had too much of a start, and the ship was destroyed. Again, one of the *Constellation*'s own had proven his gallantry.

Having already seen as much adventure in two years as many military men see in a career, Cochrane continued in the service and became the archetypal Marine officer of the late 19th century. He served conspicuously in many expeditions and engagements over the next 30 years. He rose to the rank of colonel during the Spanish-American War. Once considered for the position of commandant of the Marine Corps, Cochrane retired with the rank of brigadier general in 1905 and died in Chester, PA, in 1913.

The receiving ship *Constellation* was decommissioned 23 months after the fire on the *New Ironsides*. Her log books were then transferred to the receiving ship *Potomac*. Thus ended her days as a warship on the prowl for the enemy. Never again would the *Constellation* sail into hostile waters in search of prey. No longer suitable for combat operations due to rapid advances in technology during the Civil War years, she would soon begin an odyssey of more than one hundred years as a vessel for teaching and training.

Three

LEARNING THE STARS

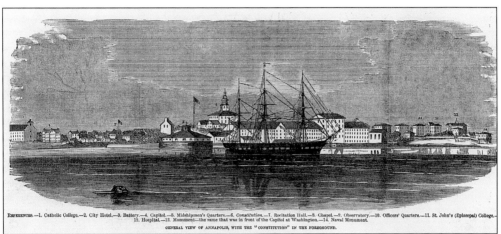

REFERENCES.—1. Catholic College.—2. City Hotel.—3. Battery.—4. Capitol.—5. Midshipmen's Quarters.—6. *Constitution.*—7. Recitation Hall.—8. Chapel.—9. Observatory.—10. Officers' Quarters.—11. St. John's (Episcopal) College.—12. Hospital.—13. Monument—the same that was in front of the Capitol at Washington.—14. Naval Monument.
GENERAL VIEW OF ANNAPOLIS, WITH THE "CONSTITUTION" IN THE FOREGROUND.

The close proximity of the United States Naval Academy to the Confederacy and the threat of a Confederate capture of the USS *Constitution*, which was being used as a school ship, forced the move of the school to Newport, RI, for the duration of the war. Despite efforts to keep the academy north, or to subdivide it, Secretary of the Navy Welles was determined that it would return to Annapolis, which it did shortly after the Confederacy's collapse in April 1865.

Former Navy Secretary George Bancroft had taken over abandoned Fort Severn from the army to begin the academy in 1845. A strong argument for the use of this former military property was the proposition that the protected harbor was ideal for maintaining a practice ship on the premises to augment classroom instruction. A strong faculty allowed the school to develop an outstanding reputation and it was soon turning out naval officers who further added to Navy laurels.

The above engraving of antebellum Annapolis reveals the early buildings of the Naval Academy as well as the famous USS *Constitution* (Old Ironsides), assigned to the school as a practice ship.

Upon the academy's return to Annapolis, Welles appointed Rear Adm. David Dixon Porter (pictured) as its sixth superintendent. Using the same tenacity he displayed during the war, Porter expanded the college buildings as well as the course offerings. By 1869, when he moved on to further his career as a full admiral, he had left an indelible mark upon the academy particularly with his installation of an honor code.

To guarantee that the Naval Academy would remain at Annapolis, Welles pushed forward the construction of several new buildings. The "New Quarters" (demolished in 1893) can be seen at the far right of this photograph taken from the north shore of the Severn River. This view offers an extremely rare image of the post-war academy. At left, dominating this picture, sits the *Constellation* with the *Essex* to her right and the hull of the *Santee* to her left. The *Santee* would serve as a floating classroom and gave the name "Santee Dock" to this area.

The Civil War monitor USS *Passaic* is captured in this image taken from the cupola of the New Quarters dome in the 1870s. Through the use of both wooden and ironclad ships, midshipmen were trained for all contingencies of naval service. The *Passaic* remained in use at the academy until 1897 when she was demolished and sold for scrap.

48

In another view taken *c.* 1870 from the New Quarters, the USS *Constitution*, soon to be replaced by the USS *Constellation*, is one of the three ships in the distance across the grounds that compose the United States Naval Academy.

To replace Porter, Commodore John L. Worden, famed commander of the USS *Monitor* at Hampton Roads, was assigned. Still bearing powder burns on his face from a direct hit from the CSS *Virginia* (formerly the USS *Merrimac*), Worden took a more conservative stance in his leadership. Following recovery from his wounds in combat, this man who was credited as the "hero of Hampton Roads," was reassigned as captain of the monitor USS *Montauk* in January of 1863. During his tenure as captain, the *Montauk* managed to destroy the CSS *Rattlesnake* in February 1863. Due to medical problems, Worden was removed from the *Montauk*, but was later returned to command in charge of the USS *Chimo*.

This close up view of the Santee Dock shows the training ships *Essex* (left), *Constellation* (center), and *Santee* (right). In 1871, the *Constellation* replaced the *Constitution* as a Naval Academy training ship. Although no longer effective for modern warfare, wooden combat vessels were maintained by the Navy well into the 20th century for training purposes. Obsolete ships like the *Constellation* and the *Portsmouth* were assigned to the Naval Academy and the Naval Training Stations.

These midshipmen are wearing the revised uniform of the late 1860s with the low stand up collar bearing an anchor on either side and the low cut kepi with gold rope lace. They are entering one of the practice ships would could very well be the *Constellation*. As these young men board with enthusiasm carrying their seabags, the regular Navy sailors can be seen along the hammock rails wondering what is in store for them in preparing these future officers.

This image, taken in the 1870s, is probably the best illustration of the *Constellation's* Civil War appearance. One exception is the overlarge gun port that has been added to the center to accommodate a 100-pound gun. This experimental endeavor was done to give the young officers a chance to train on larger caliber guns.

Drills of every description were carried out continually since Superintendent Worden believed practical instruction to be far superior to the classroom. The workday began at 4 bells (5:30 a.m.) and ended with hammocks piped down at 7 bells (7:30 p.m.) with all in bed for a much needed rest by 8. Still, the day could be extended by surprise calls to "general quarters," at which time the men were summoned out of their hammocks to their gun stations with the promise of a reward to the crew of the first gun fired.

Worden enjoyed overseeing the "cadets" (an Army term that was tagged to the mids during the 1870s, much to their chagrin) during their time aboard ship. He favored upper classmen by allowing them to use tobacco. Since most tobacco use constituted chewing, numerous spit kids or spittoons had to be placed around the decks. This prompted the captain of the ship, Jeffers, to complain that his sense of propriety was outraged for the main deck of the *Constellation* "...presented to the casual visitor the appearance of a lager-beer saloon."

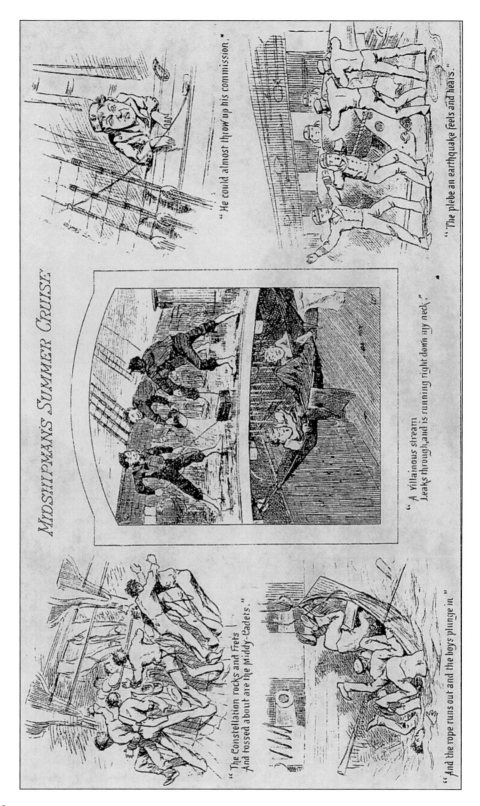

MIDSHIPMAN'S SUMMER CRUISE

"He could almost throw up his commission."

"The plebe an earthquake feels and hears."

"A villainous stream
Leaks through, and is running right down my neck."

"The Constellation rocks and frets
And tossed about are the Middy-Cadets."

"And the rope runs out and the boys plunge in"

E BELFRY
U.S.S CONSTELLATION
BOARD OF VISITORS HOUSE
THE ARMORY
OLD CADET QUARTERS
CIGAR TORPEDO BOAT
MONITOR AMPHYTRITE

True to Worden's emphasis upon "hands on" instruction, the highlight of each midshipman's year was the summer cruise where classroom lessons were augmented with real nautical experience. As the ship would slowly cruise down the smooth course of the Severn River, it gave the false illusion to the plebes aboard that life at sea was not the hardship they had been told as they believed themselves to be gaining their "sea-legs" quite easily. This erroneous belief was drastically ended when each cruise captain would have the ship heave to where the bay met the ocean to allow the ship to roll violently with the combination of the sea and the wind. Soon the rails were lined with young seasick boys and the harsh reality of life at sea was vividly ingrained upon all on board.

The illustrations on the page opposite, taken from an 1878 edition of *Fag Ends* (a combination literary, school journal, and autograph book), excellently summarize a midshipman's cruise aboard the *Constellation*. The cruise would generally take the young men from Annapolis to Newport, RI, and back. On the way they would learn sail handling, boat drill, gunnery, and navigation skills that would be necessary for them to run their own ships in the future. Of course, their arrogance as officers was tempered by their being required to also carry out the mundane shipboard cleaning routines that the enlisted men generally performed. It was hoped that such humbling tasks put upon these future officers would give them greater insight as to conditions at sea and to gain a greater respect for the enlisted man who would have to serve in these areas. To this day, midshipmen still serve a time on cruise where they function as enlisted personnel to temper their dealings with their future subordinates.

Once arrived at Newport, some classroom instruction would resume with occasional dances arranged with the ladies from local finishing schools. These dances helped to develop the social skills required of a true diplomatic naval officer.

Two entries from a "Naval Academy Alphabet" from *Fag Ends* relate directly to these teaching skills of plebes both at work and play, "...D stands for Drill, coming once every day, Which, though it looks pretty, is surely no play...N stands for Newport, the joy of each cruise, Where we waltz with the "Spoons," and they never refuse..."

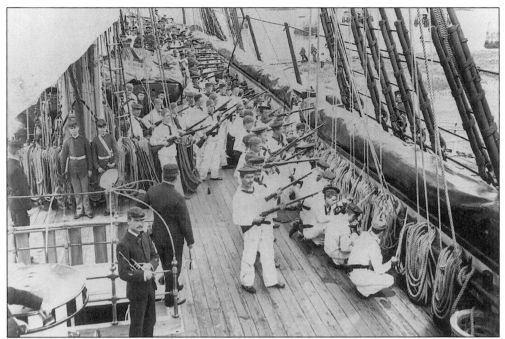

Interspersed among the midshipmen are regular Navy sailors instructing their charges on the *Constellation*'s spar deck in how to repel boarders. The mids crouched before the hammock rails are armed with cutlasses. In the second photograph the men with cutlasses have moved to the attack while the riflemen take their aim. Two marine musicians stand forward of the main mast and an instructor in the foreground clutches a trumpet which is most likely used to signal actions to be taken. These exercises were conducted at dockside with instructors appearing both on and off the ship to observe and offer commentary. The photographs were taken in the late 1880s.

This photo, part of a stereo print taken aboard the *Santee*, shows further cutlass action in another "repel boarders" exercise. Drawing largely upon their Civil War experience, the Naval Academy instructors supervised and added an element of realism to these war games. In spite of modern innovations such as the steel ship and more accurate and advanced cartridge firing guns, cutlass drill was taught to Navy personnel even beyond the years of World War I. These young men are enjoying play acting at an exercise that they all prayed they never had to use in actuality.

After a rigorous day of classwork and ship drills, two upper classmen show off for the camera by flaunting the forbidden habits of smoking and card playing. Again from *Fag Ends*, "T stands for shut off from their cruise; Also for Tobacco, which we cannot use." The new revised kepis they are wearing contrast with their Civil War surplus peacoats.

Instruction aboard ship was enriched by the Civil War experience of the instructor captains in charge. Capt. Samuel P. Carter (1819–1891) was the commandant of midshipmen from 1870 to 1873 and also commanded the *Constellation* during the practice cruise. During the war, Carter was ordered detached from the Navy to serve the Army in the western theater. He took charge of troops from Kentucky and Tennessee and rose in rank from colonel to brigadier general. His meritorious service resulted in his promotion to major general by March 1865. Following the war, he returned to the Navy, where he received promotion to rear admiral in 1882. Consequently, he is the only man in American military history to wear stars in both the U.S. Army and Navy.

Another Civil War veteran was Capt. John Lee Davis, who also later became an admiral and supervised the *Constellation*'s cruises from 1877 to 1880. Davis had formerly commanded the USS *Montauk* and the USS *Sassacus* during the war. The expertise of the veterans was invaluable to preparing young men for the possibility of action in combat.

LIEUTENANT-COMMANDER (NOW REAR-ADMIRAL) JOHN LEE DAVIS.

This 1879 image displays the most incongruous sight of the USS *Dale* with sails fully set with the exception of the spanker or driver of the mizzenmast while still secured at dockside. The crowd of midshipmen on the spar deck reveals that this is a class on setting sails. A 16-gun sloop, the *Dale* was built in 1839 and served with the *Constellation* as a floating classroom at the Naval Academy in the 1870s and 1880s. She was removed from service and sold in the early 1920s.

On March 13, 1878, the *Constellation* was relieved of training duty temporarily as Capt. James A. Greer sailed the ship to New York. There she loaded American exhibits that were to be displayed at the Paris Exhibition that year. Accompanying her was the USS *Constitution*. These two relics of America's sailing past took their cargo to LeHavre, France, after which the *Constellation* returned to Annapolis to resume her duties as a training ship. Captain Greer was yet another Civil War veteran. As a young lieutenant, he led one of the boarding parties in the *"Trent* Affair" to search for Confederate ministers Mason and Slidell. Later he commanded the ironclad USS *Carondelet* followed by command of the gun boat *Benton*. Greer became commander of the 4th District of the Mississippi Squadron, and participated in the capture of the CSS *Jeff Davis*, which entitled him to prize money. He was later selected to serve on Admiral Lee's staff, but declined in order to remain in charge of his ship.

Two years later, the *Constellation* was given another respite by serving to assist Ireland during another famine crisis. She took on cargo collected through the *New York Herald's* Irish Relief Fund. She was packed with potatoes "up to her gun deck." On April 20, she arrived at Queenstown, Ireland, where she was joined by British relief ships to complete this humanitarian enterprise. Once again a ship of war had come to serve a mission of peace.

At Queenstown the ship's officers and visiting dignitaries, including the chief financier of the expedition, the Duke of Edinburgh (front row with gold laced fore and aft cap), were photographed aboard the *Constellation*. Note the addition of a "bridge" located forward of her mizzen mast boldly displaying her name. These officers seem quite composed aboard a ship that the sailors and midshipmen complained "reeked of potatoes."

Wearing a uniform hardly changed from the Civil War era, Seaman Dunk poses in 1880 before a modern steel gun that has been placed on the *Constellation's* spar deck. Notice that the gun is not bolted to the deck. Numerous enlisted men served behind the scenes maintaining the ship and enjoying temporary authority in training boys who would soon become their superior officers.

While returning from Ireland, the *Constellation* passed numerous ships carrying immigrants to America. Thus, these new Americans were led to their destiny by this historic vessel. Her sails are being let out while a formation is established on deck. Marines at attention can just barely be discerned by their rifle barrels appearing at "shoulder arms."

In 1882, Congress passed a Personnel Act that relieved the glut of officers in the then small Navy by allowing diplomas to be awarded followed by a year of severance pay. This was also the time that the hated term "naval cadet" came into use. The only good that came of the law was the provision that academy graduates could opt to become officers in the Marine Corps. In this 1886 view, midshipmen are at ease on the spar deck during one of the *Constellation's* last training cruises. For some of these mids, a choice of the Marine Corps would be a more lucrative chance at a career than that in the Navy.

The year 1893 would be the last for the *Constellation* at Annapolis. In 1894 she was replaced by a miniature training cruiser named the *Bancroft*. These pictures capture the last days of training aboard the *Constellation* for the midshipmen. Above, the mids practice reefing a sail. Note that the spar has been lowered close to the deck to facilitate training at a safe height. Then as now, training at the Naval Academy was tiring. Below, a mid "takes a caulk" during a lull in training. A sailor sleeping on deck would often have his uniform stained with stripes from the warm tar oozing from the caulked deck seams—hence the phrase.

Lowering the ship's boats was yet another dangerous practice drill requiring skill and repetition. Here is another view of the bridge that was added to the ship in use by the officer supervising the drill. The relative calm of an anchored ship could be deceiving to anyone who would let his guard down for even a moment. One captain, while moving to observe a brilliant moon, accidentally went overboard. The marine sentry dutifully yelled, "man overboard!" An embarrassed captain yelled back from the water, "You lie, ye lubber, it's the captain." Quietly and discreetly the captain was brought back on board.

"Climb up there you monkeys!" was a common phrase used to get men aloft to cast loose or reef the sails. The agility and recklessness of youth allows these midshipmen to repeat lessons learned in safety on the deck below to where it counts—in the tops. Other mids seem to be taking a breather as they stand on the "fighting top" hanging onto the rigging. These platforms on the masts often contained Marine sharpshooters trained as snipers to neutralize officers and men on the enemy's spar deck.

A class in sail handling while at sea aboard the USS *Constellation* is illustrated here. The sails are partially reefed with some obvious flaws as uniformity is lacking. Such drills, though no doubt a welcome respite to rigid classroom instruction, were tiring and tedious as they were repeated innumerable times until they were done correctly. No doubt, some verbal "chewing out" during the exercise is producing an indelible experience among future officers.

The ship's crew c. 1890 stands at ease to the rear of the main mast for this posed photograph. The noticeable change in the uniform is the wider and stiffer flat hat. This crew would transfer the *Constellation* away from training officers in Annapolis to training enlisted men in Newport as she returned to service in the regular Navy.

After months of rigorous training aboard ship, it is time for the midshipmen to celebrate. A common practice aboard sailing ships was to have the men "man the yards," as a way of showing off their seamanship. Since there are no sails on the spars in this image, these midshipmen are certainly showing off their acquired skills learned aboard the *Constellation*. The ship would depart from the academy in 1894 after proudly serving young men who would soon defend their country commanding ships in coming wars with Spain and Germany.

Four

FOLLOWING THE STARS

On May 22, 1894, the *Constellation* was towed into the United States Naval Training Station (NTS), at Newport, RI. Other than her occasional removal for repairs and public relations missions, she would remain at Newport for over 50 years, until October 1946.

A colonial seaport town, Newport enjoyed a longstanding tradition and relationship with the United States Navy. One of the first five ships of the Continental Navy, the *Katy* (later renamed the *Providence*) was part of the Rhode Island Navy and became John Paul Jones' first command. Esek Hopkins, commander-in-chief of the Fleet of the United Colonies, was a Rhode Islander. The United States Naval Academy was temporarily located in Newport for the duration of the Civil War. By 1894, there were three naval installations in Newport: the Naval Torpedo Station (established 1869), the Naval Training Station (1883), and the Naval War College (1884). Above, Newport's harbor is depicted in a c. 1885 engraving.

The Constellation (far left) replaced the ship of the line USS *New Hampshire* as the primary sailing training ship for this facility. Here, she is depicted in a *c.* 1909 postcard, with the USS *Reina Mercedes* (foreground) and the USS *Cumberland*. Launched in 1887, the *Reina Mercedes* was wrested from the Spanish Navy when Santiago, Cuba, was surrendered in 1898 during the Spanish-American War. She was attached to the *Constellation* at Newport from 1908 to 1912. She then spent the next 45 years as a training vessel at the Naval Academy in Annapolis. The *Reina Mercedes* was condemned in 1957. She was sold to a Baltimore-based company and scrapped.

The Newport NTS was established by Commodore Stephen B. Luce in 1883. The next year, the Naval War College was established on the same site. Barracks B (shown here) was constructed in 1900 as the first permanent barracks of the facility. The *Constellation*'s masts can be seen at right.

The *Constellation* is pictured here in 1906, moored at her berth at the NTS. The two large buildings to the left house the Naval War College. The post commandant's house is visible beyond the *Constellation's* stern. The photo below shows her in Newport Harbor at about the same time. Note that her boats have been lowered into the water. The ship carried several small boats on davits, mostly located on both sides aft and across her stern. The davits were suspension systems used to lower the boats into the water. When the decks were cleared for combat, the boats were lowered into the water to keep them from being blasted apart in battle—the splinters becoming many hundreds of wooden projectiles that could cause unnecessary casualties. The ship would maneuver with the boats in tow, like a child's chain of toy ducks in a bathtub.

In 1904, the *Constellation* was placed in dry dock for repairs. In these two photos, she is pictured on June 20 of that year. Note the system of "props" employed to keep the ship stabilized in the dry dock. It was about this time that a frame one-story structure was constructed on the spar deck over her cargo hatch.

En route back to Newport, the *Constellation* was photographed at the New York Navy Yard. Here she is shown moored in an icy New York Harbor on December 17, 1904. Her rigging is undergoing repairs as can be seen by the coils of loose ends hanging from her foremast. A safety net has been stretched across her bowsprit to catch any falling workers such as the one shown working on the jib boom.

"The Constellation" the oldest U. S. War Ship Afloat.

Her erroneous reputation as the "oldest United States Navy warship afloat" caused her to be a curiosity to Newport vacationers and U.S. Navy recruits alike. Souvenirs, such as the postcard above, c. 1905, began making their way into homes across the country. The photograph of the ship shown below with her crew in the rigging, c. 1907, was also colorized and sold in postcard format. The new superstructure is visible in both views between the main mast and the foremast.

Over the latter part of the 19th century, training naval recruits became more regimented and formalized. The basics of training consisted of classroom instruction (above), coupled with military discipline (below).

U. S. Cadets at Attention.

*Newport R. I. Visual
Signaling*   2183

Service specialties were developed in the early 20th century, including training certain recruits as signalmen in all available forms of shipboard communication at that time: lights, flags, and the "wireless" or radio. Above, trainees are captured by the photographer in a hands-on class using signal flags to communicate semaphore from both aboard ship and from dockside. At left, signalmen operate a semaphore station on the dock at the NTS using flags as well as a mechanical semaphore device.

The Naval Training Stations were equipped with a sailing warship classroom not so much because the ships were obsolete, but because some of the skills necessary to operate these ships were timeless. Shown here are two postcards, c. 1907, of naval recruits aboard the *Constellation*. In the above view recruits are learning the various knots. Note the covered hammock nettings to the right and the ventilation slots of the galley cover, the galley exhaust stack to the left, and the belfry. Below, the recruits lounge about the spar deck for the cameraman.

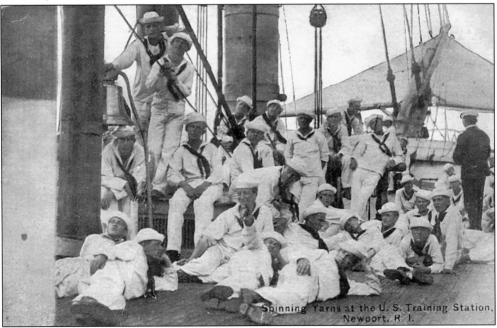

According to the caption of this 1907 view, naval recruits aboard the *Constellation* "learn the ropes." There were numerous lines or ropes that were necessary to maneuver the sails of a warship. Each line served a specific purpose and a sailor had to be able to memorize their location and use so that he could lay his hands on the correct one immediately even in the deep blackness of a moonless night.

Shown here is the earliest known photograph of the *Constellation's* gun deck, c. 1905. Its ordnance has been removed and sailors on a work detail are shown on the starboard side of the ship, with the camera facing aft toward the captain's cabin. On the right, note the capstan (used to raise and lower the anchor), the fire buckets hanging from it, and the bilge pumps.

Newport, R. I. Naval Training Station Barracks
No. 1 and 2 Newport, R. I.

Just as was done in the Army, Navy recruits were instructed in the use of small arms and battalion tactics. Here, recruits are training in front of their barracks on how to march in company formation. In the view below, the cameraman, with his back toward the ship, captures a company of recruits at "right shoulder, arms." The Naval War College is visible in the background.

U. S. Cadets. Exercising.

Unlike the Civil War period, physical fitness was becoming an issue of study and importance to the military in the late 19th century. Here, recruits are shown "exercising," according to the caption on this postcard view. Rather than mere calisthenics, it is highly likely that they are being "exercised" in the use of the rifle and bayonet in a hand-to-hand combat environment.

DRILL GROUNDS, NAVAL TRAINING STATION, NEWPORT, R. I.

In addition to training in small groups, the entire recruit brigade would also train on the drill field adjacent to the ship. Shown here in a c. 1915 view, the recruit brigade drills in battalion formation with the Naval War College in the background.

Here, in an undated post-1910 photograph, the *Constellation* puts her best foot forward. This photograph is the most detailed image available of the ship in full fighting trim. Note the complexity of the rigging system and the sails furled and secured to her yardarms. All recruits spent their time learning to fully operate the ship.

Once training was nearly completed, the recruits were ready to represent the United States Navy to the world. Here, in a 1911 postcard view, trainees participate in a parade in downtown Newport. Note the inventive mix of the dark blue uniform tops with white trousers and leggings.

Parading on the drill field with the *Constellation* in the background, recruits participate in a public ceremony. It is likely that this photograph was taken during some form of graduation program. Such formal ceremonies with their accompanying pageantry were important social occasions during this period, especially in nautically oriented, affluent communities such as Newport.

These two views, *c.* 1907, show sailors posing for a group photograph in front of and aboard the *Constellation*. Taken from a souvenir photo booklet and identified as the "Constellation Guard," they are more likely group photographs, both formal and informal, taken of a recruit company on the occasion of their graduation from the recruit training program.

Here, naval recruits are shown learning how to operate and fire ordnance known as "naval howitzers," *c.* 1908. They conduct their artillery drill on the parade ground with the *Constellation* just over their shoulder.

As was done by the midshipmen at Annapolis, the common practice in showing off a ship is done by the Navy recruits manning the yards. In this image from 1912, the familiar white stripe along the gun ports reappears.

In the first quarter of the 20th century the Electric Studio in Newport capitalized on the plethora of sailors and recruits stationed at the NTS. Personalized postcards with the sailor's photograph emblazoned on the image side were offered to veterans and neophytes alike. To the right, above, New Jersey native John Keating of the *Constellation* is shown in his blue jumper in 1913. Below, an unidentified sailor sports his "Donald Duck" hat, its ribbon announcing his assignment to the crew of "U.S.S. Constellation." The practice of wearing the name of the ship on the hat ribbon was discontinued during World War II for security reasons. The Navy deemed it unwise to advertise the identities of ships that were in port by having the sailors advertising their ships on their hats. Thereafter, these hat ribbons simply read "U.S. Navy."

In 1914, the *Constellation* was towed to Norfolk and overhauled. Once the overhaul was completed, she was ordered to Baltimore. Since she was believed to be the original frigate that was constructed in Baltimore in 1797, it seemed only appropriate to include her in the festivities that were planned to mark the centennial anniversary of the Battle of Baltimore and the defense of Fort McHenry. She arrived on September 7 and remained there until the end of October. The picture at left was taken during her stay in Baltimore and is the most commonly reproduced historical photograph of the ship. After her Baltimore visit was completed, she wintered at the Washington D.C. Navy Yard, and returned to Newport in May 1915.

Here, the *Constellation* is pictured being brought into Baltimore Harbor by two tugboats on September 7, 1914, which, ironically, was the 117th anniversary of the launch of the original frigate from nearby Fells Point.

Many activities and observances were conducted to commemorate the "Battle that Saved Baltimore." One such observance was the creation of a "human flag" by hundreds of schoolchildren holding red, white, and blue sections of cardboard. This event was resurrected years later and is presently conducted every June on the grounds of Fort McHenry National Monument.

In October 1917, the name "Constellation" was assigned to a new battle cruiser that was to be constructed at Newport News, VA. The sloop was redesignated the USS *Old Constellation* to distinguish her from the new ship, which was never completed. Construction was halted due to post-World War treaty limitations, and the ship was scrapped. Here, the cruiser's hull is shown during the scrapping process on June 7, 1924. The name "Constellation" was reassigned to the sloop in 1925.

America's entry into World War I resulted in an explosion of activity at Newport. Tens of thousands of naval recruits were trained for war duty at the NTS. Here, the brigade of recruits are shown at parade on the drill field in the summer of 1917, with the *Constellation* at the end of the pier.

The same view of the brigade of recruits is shown in the winter of 1917–18, with the now redesignated *Old Constellation* shown at her berth.

CENSORED

COPY. BY UNDERWOOD & UNDERWOOD, N.

NEAT AND TRIM, NAVAL RECRUITS AT NEWPORT LINE UP FOR INSPECTION.

After the crisis of World War I passed, Newport and the *Constellation* settled back into their role of training a peacetime Navy. Here, a postcard from 1923 illustrates recruits on parade. The *Constellation* was docked to the left of this scene off camera. New wars still required traditional discipline.

While she continued to fulfill her mission as a floating classroom, the *Constellation* also continued to serve as a landmark on the Newport waterfront. In 1933, this photograph of the Naval War College and the NTS was taken from across the harbor. She was decommissioned on June 16 of that year and surveys were conducted to determine how best to restore her. No action was taken, however.

Another unidentified member of the training ship *Constellation*'s crew is captured here by a local photographer. The pedestal and pre-tied knot props served as evidence that this photographer received a lot of business from the local naval installations. This photograph of an unidentified sailor appears to be *c.* World War I. The white lanyard around his neck holds the sailor's knife carried in his jumper pocket. Life at the Newport NTS was quite pleasant in the booming, glory days of the 1920s. Striped by the shadows cast by the rigging, the ship's officers were photographed on a bright sunny day with their pet dogs on the *Constellation*'s spar deck, shown below.

The perception of the ship as the original frigate developed a life of its own. Here, the *Constellation* is photographed on February 9, 1922, dressed out with large national colors posted on each of her masts on the 123rd anniversary of the frigate's victory over the *L'Insurgent*.

Apparently, this practice of photographing the ship as it was dressed out on the anniversary of the frigate's victory may have become somewhat of a ritual. Here, exactly two years later, recruit companies training at the NTS stand in formation dockside, with the *Constellation* as a backdrop. February 9, 1924, was significant in that it was the 125th anniversary of the frigate's victory over *L'Insurgent*. This photograph shows six companies of recruits in formation under

U.S.S. Constellation
Oldest Ship in the World
Ceremony commemorating her Victory over the
l'Insurgente 9th February 1799

arms and wearing pea coats and leggings. A color guard is also conspicuous, as are officers dressed in overcoats and armed with their ceremonial swords. The entire formation is shown at "present arms": the formal salute rendered by personnel under arms. By now, the *Constellation* was 69 years old, and the effects of time were starting to become more readily visible. Note the deterioration visible in the cutwater, hammock nettings, and bulkhead area.

In May 1926, the *Constellation* received orders to proceed to the Philadelphia Navy Yard to participate in the nation's sesquicentennial celebration in July. She is shown here in full holiday dress prior to her May 13th departure. The Naval War College is prominent on the hillside beyond.

Chief Boatswain E.D. Delazy and 28 hands nursed the *Constellation* to the birthplace of our republic, under tow by an ocean-going tugboat. She arrived at the Philadelphia Navy Yard on May 15, and remained past the celebration until late November, arriving back at Newport on December 1. Here, she is shown at the navy yard with an early submarine in the foreground.

If a nation can possibly have a birthplace, America's is Philadelphia. Therefore, "the City of Brotherly Love" was chosen to host the centennial anniversary celebration of the Declaration of Independence in 1876. Fairmount Park was chosen as the location for a great exposition commemorating the event. Although the celebration was shrouded in shock and disbelief at the news of the death of Civil War hero Brevet Maj. Gen. George A. Custer in battle against the Sioux and Cheyenne tribes in Montana in late June, the exposition was an overwheming success.

Fifty years later, Philadelphia was again chosen to host the national sesqui-centennial celebration in 1926. At the height of the "Roaring Twenties" economic boom, $26 million was spent on another exposition that was held on the current location of Franklin Roosevelt Park and the Veterans' Stadium complex. Although a critical success, it was a financial disaster as the event drew only six million visitors.

Here, the *Constellation* awaits visitors at the nearby Philadelphia Navy Yard during the celebration, serving as part of the Navy's contribution to the festivities. To the left is the bow of the cruiser USS *Olympia*. She was the flagship on which Adm. George Dewey uttered the famous order "You may fire when ready, Gridley" which opened the Battle of Manila Bay in the Spanish-American War. The *Olympia* is the last remaining ship afloat that saw combat in the "Splendid Little War" of 1898. She is open to visitation at Penn's Landing in downtown Philadelphia.

On August 24, 1940, the *Constellation* was placed back into commission by order of the President of the United States. Part of a commissioning ceremony included flying a narrow commissioning pennant from the top of the mainmast. Pictured here is the pennant used by the *Constellation* upon her recommissioning.

As a novelty of days gone by, the ship continued to be an object of curiosity. On May 8, 1941, the national commander of the American Legion and his staff visited the ship. Here, they pose with two of the ship's officers. Note the sailor on the right assuring his place in the picture.

Days later, the *Constellation* was designated relief flagship of Adm. Ernest J. King (1878–1956), commanding officer of the Atlantic Fleet, largely a *pro forma* honor. A 1901 graduate of the Naval Academy, he served on the staff of Adm. H.T. Mayo, commander of the Atlantic Fleet during World War I, having by then advanced to the rank of commander. On December 30, 1941, Admiral King became the commander in chief of the U.S. Fleet, coordinating all U.S. Naval forces during World War II. In December 1944, King was promoted to fleet admiral, the Navy's equivalent of a "five star general." Here, King is shown inspecting the USS *Boise* at the Philadelphia Navy Yard after she was damaged in the Solomon Islands Campaign. He is escorted by Capt. Edward J. Moran (left), skipper of the *Boise*. Rear Adm. Milo F. Draemal walks behind King.

For her new mission as a flagship, a frame building was again constructed on her spar deck and the gun ports were enclosed with windows. This photograph was snapped in September 1941, ten weeks prior to the Pearl Harbor disaster. As relief flagship of the Atlantic Fleet, the first word of the calamity reached the East Coast via the U.S. Navy signal service on the *Constellation's* ship's radio, believed to be contained in one of the added structures.

Here, the crew of the *Constellation* poses for a group photograph in front of the ship on November 8, 1941, 29 days before the Japanese attack on Pearl Harbor. The row of seated men consists of the ship's two commissioned and eight non-commissioned officers. The rest of the crew of 75 men pose in several rows behind them. Note the conscious effort to ensure the "1797" plaque is visible in the photograph.

After Admiral King was promoted to naval chief of staff, Adm. Royal E. Ingersoll (1883–1976) employed the *Constellation* as his relief flagship. A second generation Naval Academy graduate, he was the son of Rear Adm. Royal R. Ingersoll, a veteran of the Civil War and later a senior staff officer in Theodore Roosevelt's "Great White Fleet." Much of the father's career was spent as a professor at the Naval Academy. The younger Ingersoll (left) was a Naval Academy classmate and friend of Adm. Chester Nimitz.

Luncheon to Admiral R.E. Ingersoll, Admirals Cabin On USS Constellation on his 60th Birthday June 20 1942 to [illegible]

On June 20, 1942, Ingersoll was photographed in his quarters aboard the *Constellation*, celebrating his birthday by hosting a dinner party for his staff. Rear Adm. W.K. Kilpatrick is seated to the Admiral's right and LCDR John Davis, the captain of the *Constellation*, is to his left.

America's isolationist tendencies disappeared on December 8, 1941, when the United States declared war on Japan. Sixty-four days later, a letter left the ship in this patriotic envelope with postage that promoted the national defense. The eagle logo and lettering is embossed in gold.

In August, 1942, movie star Tyrone Power enlisted in the U.S. Marine Corps. His basic training was delayed so that he could star as a naval officer in the wartime movie *Crash Dive*, filmed in part in Newport. Here, Power is pictured with the *Constellation's* captain, Lt. Comdr. John Davis. The ship appears briefly in the background of the movie. Power reported for duty upon completion of filming in December. Already a licensed pilot, he was sent to the officers' school in Quantico, VA, and was a first lieutenant when discharged in January 1946.

In 1942, Ingersoll (seated third from right) poses for a photograph on the *Constellation's* spar deck with his staff consisting of 12 naval officers and one marine officer. The original of this photo was autographed by all of the officers pictured. Ingersoll is flanked by Rear Adm. Olaf M. Hustvedt and Col. William E. Riley, USMC.

Newport was no exception to the mobilization for the war effort. Just as she did in the "War to End All Wars," Newport continued to be a principal training facility for naval recruits and draftees. Here, recruits engage in the age-old chore of "holystoning" the *Constellation's* decks, a practice that dates from the earliest days of the Navy. Holystones were blocks of pumice roughly the size of a common bible that were used to scour the decks, usually conducted by sailors on their knees, thus generating the "holystone" moniker. These recruits are shown on the berth deck, barefooted with their trouser legs rolled up, applying the holystone with a handle.

Among the many activities to which recruits and draftees were introduced was the seabag inspection. A sailor only brought aboard ship what he could fit into his seabag. This photo was part of a souvenir booklet sold at Newport during this period.

Even this relic of the pre-Civil War sailing Navy served its role in a war that saw the dawn of the nuclear age. Here, recruits stand in formation in front of the *Constellation*, with the Naval War College again visible in the background.

When the war in Europe ended, Japan remained the only hold out. Here Lieutenant Sabitsky assumes command of the *Constellation* from 67-year-old Lt. Cmdr. John W. Davis on May 31, 1945. Davis, who was called out of retirement to command the ship for the duration of the war, received the Medal of Honor for heroism in Cuba during the Spanish-American War.

Five

THE STARS REALIGNED

With the end of World War II also came the end of the *Constellation*'s naval service. She was towed to the Charlestown Navy Yard in Boston and there allowed to languish and rot at her moorings. Her sad state of affairs stood in stark contrast to the well-kept USS *Constitution*, being maintained by the U.S. Navy, directly adjacent to her. Studies of these two ships side by side vividly reveal the difference between the 18th- and 19th-century configurations of the two vessels, particularly in the bow and stern design.

Due to the efforts of a few patriotically minded sponsors, the *Constellation* was saved from destruction, and in 1949 Congress passed a bill to restore her if 75 percent of the $4,525,000 cost could be raised by the public. The public effort failed and soon the *Constellation* was deteriorating at a more rapid pace. Navy requests to destroy the ship were rejected by the Congressional committee, and the cause to bring her to Baltimore was picked up by the private non-profit Star Spangled Banner Flag House Association. A 1954 bill passed Congress authorizing the ship to come to Baltimore, where this association would take over the restoration. In August of 1955, the *Constellation* was towed onto a floating dry dock for the passage to Baltimore, MD.

The Constellation is shown above being maneuvered into the U.S. Navy floating dry dock *ARD-16*, in preparation for her move to Baltimore. Narrowly escaping a seasonal hurricane, the *ARD-16* delivered the *Constellation* safely into the hands of the good people of Baltimore. Below, the ship is seen passing under the newly completed Chesapeake Bay Bridge aboard the *ARD-16*.

With the help of tugboats, the *Constellation* moves through the gates of the Pennington Avenue drawbridge to a temporary berth, after being freed from the *ARD-16*.

She was taken to her mooring at Pier 4, Pratt Street, where her care was overseen by the newly formed Constellation Committee of Maryland. Now came the difficult period of raising funds to restore the ship and pay for a permanent berth. The next nine years were hard and lean times for the ship as funds were slowly raised. During this time she fell into even greater ruin as evidenced through the following pictures.

Falling victim to further dry rot, vandalism, and rat infestation, it seemed restoration was a hopeless dream. A wooden ship generally rots from the top down due to the invasive properties of rain water, whereas salt water tends to serve somewhat as a preservative. The *Constellation* had never looked so dilapidated in all her one hundred years of existence. The inside of the ship was even worse.

While awaiting restoration most of the caulking on her forepost was lost due to dry rot and severe weathering. To further complicate matters, the old argument as to her origin as a frigate or as a sloop re-emerged. To best tie her in with the city of Baltimore, the restorers wanted to remake her as the original frigate. Not only did this belie her present lines but it began a paper battle between historians that put the ship squarely in the middle of controversy that only served to slow fund-raising.

By 1964, those desiring her to be put into her 1812 appearance won out and the ship was sent to the Maryland Shipbuilding & Drydock Company for her first restoration. In this photograph, her past history is reflected from her Civil War–era hammock rails to the World War II radio sheds on her spar deck. This would be the last appearance of the sloop of war *Constellation* for 35 years as well meaning but misguided efforts were made to make her appear like her frigate namesake. Fortunately, the restoration left her hull intact, thus saving most of her original fabric.

Massive supports and scaffolding soon surrounded the ship as repairs began. Since her hull was in relatively good shape, her superstructure received most of the attention. Concern for the ship was great among the many workers at Maryland Drydock. This concern coupled with patriotic zeal resulted in many employees contributing additional off-hours' time to the completion of her restoration.

The bowsprit and hammock rails were removed so that the ship could be refitted as a frigate. The entire superstructure was based on the 1797 configuration and was merely added on top of her true 1854 hull. Solid bulwarks with cutouts for spar deck carronades were added.

All of the World War II additions were removed with the exception of a radio station that was kept below on the berth deck. A new bowsprit was added as well as new masts and spars. New hammock irons were made to line the newly added bulwarks. Her transformation was a remarkable piece of construction.

By 1971 her supports were removed and she was refloated and repainted in preparation for her new rigging. Her new bulwarks can be seen in this view. Seven years of earnest restoration now returned the ship to the city of Baltimore.

Although historically incorrect, the *Constellation* sits proudly at her moorings with her guns poking out of her ports making her appearance as the "frigate." Her rounded stern has been fully removed and replaced with the flat stern that has been handsomely decorated. A hand carved eagle made by a local Baltimore volunteer also adorns her stern transom.

This overhead view shows the magnificent restoration effort made from the 1960s through the 1970s. The large open area between the fore and main mast is known as the ship's well. Although an integral part of a frigate, it was unknown to the workers that this opening would allow the keel to begin bending upward creating a slowly growing "hog" to the ship's structure that would take years to detect. The well had to be covered when visitors were aboard to prevent children and less observant persons from falling through to the gun deck.

At this point, Baltimoreans accepted that the ship was the original frigate launched from their city in 1797. Regardless of this error, visitation increased as did civic pride in this vessel of war that was now becoming a fixture and well-known landmark within the inner harbor.

In the days before heavy insurance litigation, tugboat pilots took great pride in being asked to move the *Constellation* in and out of the harbor for turning to allow her to weather evenly on both sides.

Interior repairs were also being made to the ship as well. Photographs below decks are quite rare but here is one showing the officers' wardrooms with the doors open. Due to the spindle covered ventilation ports, many visitors mistakenly believe these officers' living areas to be prison cells or the brig.

In great need of repair is one of the officer's bunks. Note how the drawers for storage are located beneath the bed proper. The rails are strategically located to keep the occupant from pitching over the side in a rough sea. Also note that the port holes have disappeared. There would be at least one port located here for the ventilation of the berth deck.

To properly display the vessel as a warship, guns were an absolute necessity. These 18-pounder replicas were made of fiberglass to give a correct appearance while cutting expense as well as unnecessary weight. Two authentic cast iron guns were displayed at the middle gun ports of the ship on each side.

In 1976, the hull needed to be repaired and the ship was returned to Maryland Drydock. This would be the first of many return trips for various repairs and upgrades. Copper sheathing was added as per the original to stop infestation of freshwater worms and to inhibit the growth of marine life along the hull. Originally a well-coppered hull would also aid in adding speed to the vessel.

With her rigging once again removed, the *Constellation* nevertheless provides a magnificent sight proudly illumined in this night shot at dry dock. Now that she had become a regular fixture in the Inner Harbor, donations were coming in more frequently which enabled The Constellation Foundation to maintain her better. Here, she sits in the Fort McHenry shipyards of the Bethlehem Steel Corporation.

This remarkable shot of a fully rigged ship totally out of the water was taken at the Bethlehem Steel shipyard. Only the props holding her stable are in place as all the scaffolding has been removed. She sits waiting for the flood gates to open which will allow her to be refloated in preparation for her return to the inner harbor. Once the dock is flooded, the water will stabilize the ship and the props can be safely removed.

Repairs complete, the *Constellation* floats out of dry dock in all her full raiment as an 18th-century frigate to return to her berth in 1979.

During these many refittings, the Constellation Committee had changed its name to the US Frigate Constellation Foundation. This enterprising group had coins struck from the discarded original copper rivets of the ship. These coins, which provide unlimited visitation to the ship, were sold to raise necessary funds. This idea was very successful and profitable and the coins are still honored today in spite of the misinformation that is labeled on the coins. Many Baltimoreans treasure these coins as a proud reminder of civic pride and the early days of her restoration.

Celebration and fanfare were always a part of each return of the ship to generate funds and historic pride. By now, it was well accepted among Baltimoreans that this ship was a great advertisement for a city that was trying to rebuild itself. Just as the ship had been restored, the city was striving to remodel itself to attract visitors from near and far.

The Continental Marine Corps, forerunner of the U.S. Marine Corps, was established on November 10, 1775. This anniversary continues to be celebrated annually. Since the first Marine officer to receive his commission on board a ship happened on the frigate *Constellation*, it became a ritual to have the annual observance held at or aboard the ship. Here former Maryland State Comptroller Louis Goldstein, a Marine Corps veteran of World War II, officiates at one of these celebrations.

In the late 1970s, The Rouse Company was granted a contract to renew the Inner Harbor by building a development to be called Harborplace. The *Constellation* was to be the centerpiece for this landmark effort to revitalize the city. While construction begins for this new enterprise, the ship awaits her new surroundings.

113

The new construction would call for two huge pavilions to be built. One would be placed just forward of the ship to the north containing retail establishments. An amphitheater for spectator entertainment would appear approximately where the circular walkway appears in this photograph. Finally a second pavilion for specialty food services would be placed on the far left. The *Constellation* would then be centrally located among the improvements.

Pier one would now be forever known as Constellation Dock now that the ship had secured her permanent berth. Here tugboats ease the *Constellation* into her place of honor.

When Harborplace opened in 1980, the ship was firmly established as the centerpiece. Due to the success of this revitalization effort, visitation rapidly rose from a few thousand per year to sometimes over a thousand in a day. Soon other cities vied to emulate Baltimore's success despite the fact that they lacked an attraction as colorful and unique as the *Constellation*.

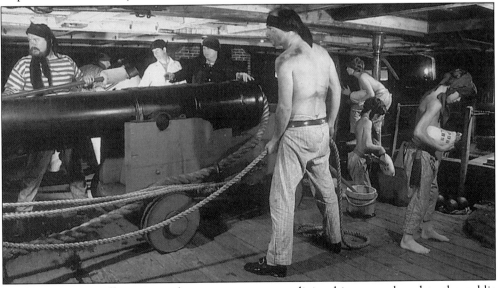

In 1981, a serious effort was made to present accurate living history on board to the public through the volunteers of Ship's Company, 1800. To accompany the representation of the ship as a frigate, the group performed gun drills c. 1799–1800. Despite the arguments over the ship's authenticity as a frigate, her reputation throughout the museum ship community was enhanced by this group's sharing of its research and its introduction of naval living history to the USS *Constitution*, the brig *Niagara* in Erie, PA, and the HMS *Victory* in Portsmouth, England.

With the resounding success of Harborplace, which attracted millions of visitors worldwide to Baltimore, new attractions were added. The submarine USS *Torsk* and the U.S. Coast Guard cutter USCGC *Roger B. Taney*, both veterans of World War II, would augment the ship. The National Aquarium in the background drew additional visitation to a remodeled Baltimore.

Controversy in regard to the ship's history was overshadowed with the addition of the new visitor's center on Constellation Dock. Although such a facility was greatly needed, complaints rang out that the building obstructed the view of the ship and was an eyesore in the middle of Harborplace. The *Constellation's* board of directors successfully fended off the criticism but more pressing problems were on the horizon.

The ship passed through the 1980s variously used for historic interpretation and entertainment, as well as military and civic ceremonies. Regardless of political change within the city, the ship maintained its status as an icon of the Inner Harbor. However, discerning eyes began to notice that the ship seemed to be bulging up in the middle or "hogging" as a seaman might say. Despite being formally out of commission by the U.S. Navy since 1955, a naval inspection team came on board to investigate the apparent problem. What they found was absolute disaster. Not only was the ship bulging up to the point of breaking, her timbers were no longer able to support her superstructure. Removal of the masts failed to correct the problem and in 1993 the ship was officially condemned by the Navy and all visitation was ended. Now debate began as to what to do with a ship that was virtually useless and slowly destroying itself. Nylon bands were strapped around the body of the ship in a feeble attempt to try to hold her together. To add insult to injury, the old arguments as to her true origin resurfaced and critics urged that she be taken out to sea to be sunk with a last ceremonial effort of dignity.

But the Constellation Foundation assisted by interested public and private figures were not ready to give up this ship. Soon a letter writing campaign led to government hearings, particularly at the state level. With a strong campaign slogan of, "Some things are worth fighting for!," supporters of the ship won a reprieve. A unique plan submitted by naval architect Peter Boudreau, architect of the *Pride of Baltimore II*, was accepted by both the U.S. Navy and the State of Maryland to save the ship. At the heart of the plan was the aim to restore the ship at last to her rightful configuration as an 1855 sloop of war. In November of 1996, the heavily leaking hull of the *Constellation* was towed from her berth to dry dock at Pier 5 near Fort McHenry. A project estimated to cost $9.5 million, paid by a combination of local and state grants matched by corporate and private donations, was begun.

By April of 1997, all of the false superstructure was gone and the ship was reduced to the deep black timbers of her original fabric. She appeared much like a skeleton of some great beached whale with but little resemblance to her former glory. However, a good 30 percent of her original construction still remained and she was in better shape than at first felt. By June of 1997, the original timbers were stabilized and new construction was begun.

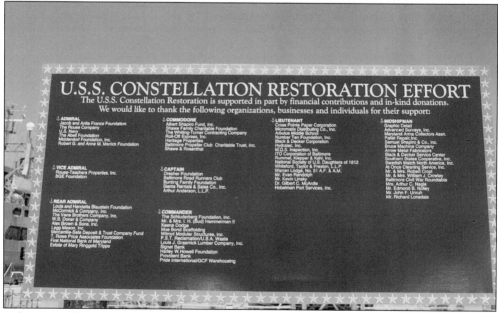

A large sign listing contributors to the project was placed near Fort Avenue to both thank donors and advertise the ship which was open to visitation from the public. The good will generated by these "open houses," as well as a chance to see the ship construction at first hand, created strong interest in the project and gained more donations along with much needed volunteer help.

A laser survey of the hull compared to John Lenthall's original plans verified beyond a doubt that this was the 1854 sloop. To further add to the archeological evidence, carpenter's marks such as this one showing a perpendicular to the keel, bore the telltale professional markings of craftsmen from the 1850s. All archeological findings were documented for future reference.

Boudreau's plan called for a "cold molding" process that utilized modern epoxies to laminate layers of Douglas fir to help form the ship while encasing the original structure. Although this process had been used successfully in the making of smaller vessels, this would be the first effort using such a process on a ship of this size. This project became the largest maritime restoration process ever undertaken by a non-governmental agency.

The lines of the ship begin to return her former grace as four thicknesses of fir planking are placed forming the sides of the hull. A layer of horizontal planking is covered with two layers of diagonal planking running in opposite directions and then covered with another layer of horizontal planking as shown here. The large openings are for the gun ports on the gun deck.

By June of 1998, the ship was fully regaining its appearance as a ship. The carpenters who gave of their time in a truly loving and caring fashion were becoming amateur historians as they studied the original plans to try their best to replicate the work of their forebears over a hundred years before. Here the bow takes on its distinctive shape.

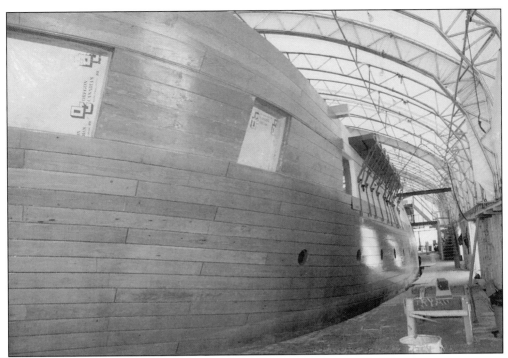

With the planking completed, the treated original chains are added to the port side as the ship begins its reappearance. The gun ports as well as the windows of the captain's cabin are now easy to define. The roofing is a huge canopy erected to keep the ship and the construction workers protected from the elements. A separate workshop area including its own blacksmithing shop was created to rebuild all of the fittings in the old style.

Once again the rounded stern of the 1850s reappears. Above this deck the folding bulwarks will be placed that enable the stern pivot gun to operate. Coats of protective epoxy seal the wood and numerous coats of paint begin to restore the luster of the old ship. Some copper will be placed to cover more sensitive areas of the original hull. All of the ship's bottom will be painted green to simulate oxidized copper.

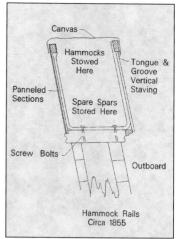

Canvas

Hammocks
Stowed
Here

Tongue &
Groove
Vertical
Staving

Panneled
Sections

Spare Spars
Stored Here

Screw Bolts

Outboard

Hammock Rails
Circa 1855

Fortunately, much of the original ironwork escaped being sold for scrap and was in the hold of the ship being used as additional ballast. Here are the original hammock rails that would eventually be returned to their original use to form the lining walls or bulwarks of the spar deck. Each day the men would roll up and tie their hammocks into a sausage like tube. The hammocks were then stowed inside the hammock rails with the man's number appearing on the upright surface of the hammock for easy identification.

August of 1998 was a momentous occasion as the completed hull structure was now ready to be refloated. Here the full majesty of the ship is revealed as she sits in dry dock awaiting the floodgates to be opened. In this view the hammock railings have yet to be installed. The sleek lines of John Lenthall's original design are readily discerned.

Fresh paint highlights the new quarter galleries located at the rear of the gun deck and indicating the location of the captain's cabin. The captain would have his own private toilet located in this area which included his bed, desk, dining area, and storage facilities. An emergency compartment used to temporarily operate the rudder should the cables be shot away, is also here. Note the worker standing at the base of the ship's rudder creates perspective of the size of this warship.

With much fanfare and ceremony, the *Constellation* was successfully refloated on August 18, 1998. All of the efforts to raise money to restore this ship now came to fruition. A project that seemed so hopeless just a few years before now showed promise of a successful future.

The success of fund-raising largely depended upon the private sector. As one example among others, Arbutus Middle School of Baltimore County, Maryland adopted the *Constellation* for a project to involve students in community history. Here the students proudly present a check for $1,316.50 to the Constellation Foundation. The funds were raised by student donations of spare change at lunch. These "pennies for *Constellation*" were reminiscent of efforts made by schoolchildren at the turn of the 20th century to save the USS *Constitution*. History has repeated itself.

Not only was the ship restored as accurately as possible to her original configuration, but ancient nautical traditions were followed as well. Middle school students helped naval architect Boudreau place coins under the base of the mast. Tradition has it that these coins would pay the boatman's fee to cross the mythical River Styx should the ship flounder. Both modern and 1850s vintage coins were placed and the ship was returned to its status as the great educator of its nation's youth.

Just one week prior to her scheduled return to her berth in the Inner Harbor, the *Constellation* began rigging operations. The crew worked round the clock to have her ready for her return on the 19th anniversary of Harborplace. Since funding only covered the structural and exterior restoration of the ship, some concessions had to be made to maintain her proper appearance as a warship. Consequently, her 18th-century reproduction 18-pounder guns were temporarily re-used to fill her open gun ports until correct Civil War period ordnance could be financed, constructed, and installed.

On July 2, 1999, the restoration team gave Baltimore an early Independence Day gift for the nation's 223rd birthday as the ship returned to her traditional place at Constellation Dock in the Inner Harbor. An enthusiastic crowd of Baltimoreans were there to welcome her home. Her new custodians, Living Classrooms Foundation, assumed responsibility to lead her in her 21st-century mission to serve again as a school ship, assisting the efforts of Maryland schools in educating its youth. Her legacy would now span three centuries.

To firmly establish her Civil War vintage, Ship's Company updated its image to the 1860s. The history of the ship comes full circle as this group, attired in Civil War–period uniforms, stands in formation on the quarter deck ready to receive visitors and to properly interpret the ship. They stand equally dedicated to assisting Living Classrooms with their mission of preparing for the future through the study of the past.

126

The *Constellation*'s homecoming brought all of the ship's history together with a melding of the past into the present. The ship was returned to its proper appearance through the hard work and persistent efforts of public figures and private citizens alike. Baltimore Mayor Kurt L. Schmoke, a key political supporter of the restoration, represented both the local community support as well as the African-American heritage of the ship while speaking to those assembled. Also present were such key supporters as Senator Paul Sarbanes, representing the federal level, and Governor Parris Glendenning, representing the state support of both himself and his predecessor, Governor William Donald Schaeffer.

The *Constellation*'s legacy lives on through the re-use of her proud name with "America's Flagship," the aircraft carrier USS *Constellation* (CV-64). Commissioned in 1961, she continues the mission entrusted to her two former namesakes to protect American interests throughout the world.

Although her restoration is an ongoing process, the USS *Constellation* sits majestically at her berth awaiting visitors. Serving as the only surviving warship afloat that actively served in the American Civil War, this ship stands ready to tell the story of the U.S. Navy's role during our country's greatest test of endurance. The story of her very existence serves as a focal point of the story of how a nation divided was restored and continues to provide hope for the future.

In the ill-fated fight of the USS *Chesapeake* against the HMS *Shannon* in the War of 1812, American Capt. James Lawrence uttered his last dying words, "Don't Give Up the Ship!" Just a few months later Commodore Oliver Hazard won the Battle of Lake Erie with all of his ships bearing flags carrying those immortal words. They appear on the cutlass rack of the USS *Constellation* to remind us today as well as it did the crews of the past of our duty to protect and preserve this historic ship. Indeed, *"Some things are worth fighting for!"*